CISM COURSES AND LECTURES

The series presents lecture notes, monographs, edited works and proceedings in the field of Mechanics, Engineering, Computer Science and Applied Mathematics.
Purpose of the series in to make known in the international scientific and technical community results obtained in some of the activities organized by CISM, the International Centre for Mechanical Sciences.

INTERNATIONAL CENTRE FOR MECHANICAL SCIENCES

COURSES AND LECTURES - No. 350

VISCO-PLASTIC BEHAVIOUR OF GEOMATERIALS

EDITED BY

N.D. CRISTESCU
UNIVERSITY OF BUCHAREST
AND
UNIVERSITY OF FLORIDA

G. GIODA
POLYTECHNIC OF MILAN

Springer-Verlag Wien GmbH

Le spese di stampa di questo volume sono in parte coperte da
contributi del Consiglio Nazionale delle Ricerche.

This volume contains 209 illustrations

In order to make this volume available as economically and as
rapidly as possible the authors' typescripts have been
reproduced in their original forms. This method unfortunately
has its typographical limitations but it is hoped that they in no
way distract the reader.

ISBN 978-3-211-82586-0   ISBN 978-3-7091-2710-0 (eBook)
DOI 10.1007/978-3-7091-2710-0

PREFACE

An increasing interest exists nowadays for the time-dependent, visco-plastic behaviour of rocks owing to its large influence in important applications of mining, civil and petroleum engineering as well as in geophysics, seismology, etc.

Aim of this work is to present to researchers and engineers working in the above mentioned fields a description of the present knowledge on the viscoplasticity of geomaterials, considering the experimental and theoretical aspects and the applications to relevant problems through numerical, computer oriented methods.

First, the laboratory tests performed in order to quantitatively assess the rheological properties of rocks are illustrated, describing the equipments used for this purposes in modern rock mechanics laboratories and the proper testing procedures.

The above experimental information forms the basis of mechanical models able to describe the time-dependent behaviour of rocks such as deviatoric creep and relaxation, as well as specific properties of geomaterials as dilatancy and/or compressibility during creep, long term damage and failure, etc. Part of the presentation is devoted to the description of constituive creep laws, and it is shown that the mentioned properties can be accounted for in the framework of constitutive models of increasing complexity.

The book covers also problems related to the application of numerical solution procedures to rock/soil mechanics situations in which the slow deformation of geomaterials is to be taken into account. Among the various numerical solution techniques, the finite element method is considered in details. In this section, in addition to the analysis of visco-plastic problems, also the time dependent behaviour induced by the interaction between solid particles and pore fluid (or consolidation) in two-phase media is discussed.

Finally, some applications are discussed, with particular reference to the design of underground openings in creeping rock, as creep closure of vertical shafts or horizontal tunnels, failure around underground openings, etc.

Nicolae Cristescu
Giancarlo Gioda

CONTENTS

Page

UNIAXIAL AND TRIAXIAL CREEP AND FAILURE TESTS ON ROCK: EXPERIMENTAL TECHNIQUE AND INTERPRETATION

U. Hunsche

**Federal Institute for Geosciences and Natural Resources (BGR),
Hannover, Germany**

ABSTRACT

Modern test equipment and experimental techniques for uniaxial and triaxial failure tests, deformation tests, and creep tests are shown with special emphasis on tests on soft rocks and on time effects. A large number of test results, mainly on rock salt, and their interpretation is given. Besides uniaxial and triaxial creep tests on cylinders, true triaxial tests on cubes and tension tests are also included. Determinaton of elastic constants, performance of stress drop and relaxation tests are presented as well. Emphasis is placed on the identification of the physical mechanisms for creep and fracture and on the determination of volume change.

1. INTRODUCTION

Knowledge of the mechanical properties of rocks is important for many engineering purposes, e.g. for the design of mines, tunnels, repositories, and deep wells, as well as for application in geophysics, tectonics, and seismology. In general, the elastic properties, creep, dilatancy, damage, fracture development, failure, and permeability are of interest for this purpose (see Langer, 1986). For corresponding model calculations one needs constitutive laws and equations describing the dependence of the properties on various parameters. In some special cases also probability density functions for the material properties have to be determined.

Most of the above mentioned properties are coupled, e.g. dilatancy and permeability. Therefore, this dependence has to be evaluated as well. It has also to be regarded that many properties are time or rate dependent, e.g. failure strength. Another difficult question is whether the material under question has a flow limit or a yield stress.

For the development of reliable laws appropriate laboratory and field tests have to be performed. In this publication the experimental techniques for the laboratory creep and failure tests and their interpretation, and tests for determining elastic properties are described. Emphasis is placed on soft rock using rock salt as example. Creep is defined here as irreversible deformation without fracturing. A limited number of constitutive equations are shown for rock salt. The knowledge about rock salt is rather far advanced and it is often used as a model material, not only in rock mechanics but also in materials sciences.

The knowledge of the experimental technique and the interpretation of tests is not only important for the experimentalist but also for the theoretician who wants to develop reliable and applicable constitutive equations since otherwise serious misinterpretations and mistakes are very likely to occur. Therefore, close cooperation between experimentalist and theoretician is an essential condition. Therefore, do never use experimental results without careful inspection of the whole test procedure.

The development of appropriate constitutive equations can be made on the basis of plasticity theory, rheological bodies, or on the basis of microscopic deformation mechanisms. Also combinations are possible. The use of purely empirical formulas should be avoided, because the great source of expert knowledge is not included. For the extrapolation of the results for long times or for experimentally not covered conditions the relevant microscopic mechanisms have absolutely to be included. At present, one has the impression, that plasticity theory and microscopic theory are often combined. The true triaxial failure tests in chapters 2.3 and 3.1 have been used by Cristescu & Hunsche (1992, 1993a,b) and by Cristescu (1994, this volume) for the development of an nonassociated elastic-viscoplastic constitutive equation for rock salt.

It is not intended to cover rock mechanics in total; this publication is focused on time dependent effects in rock mechanics. Moreover, it covers mainly the experimental technique used in the laboratory and the interpretation of tests. Some theory of creep deformation is given, theory of dilatation and failure is only touched. In addition, the contents of the publication is strongly influenced by the personal experience of the author. Therefore, the reader is encouraged to read other publications. The following general books are mentioned: Paterson (1978), Jaeger & Cook (1979), Jumikis (1983), Lliboutry (1987), Cristescu (1989), and especially for rock salt: Dreyer (1972), Hardy & Langer (1984, 1988). As sources for rock data the following books are mentioned: Lama & Vutukuri (1974/78), Clark (1966), Gevantman (1981), Landolt-Börnstein (1982), Carmichael (1982, 1984). Moreover, standards, suggested methods or recommendations for experimental procedures and their interpretation are given by several national and international commissions, e.g. ISRM (International Society for Rock Mechanics), ASTM (American Society for Testing and Materials), and DGEG (Deutsche Gesellschaft für Erd- und Grundbau = German Geotechnical Society). The reader is reminded, however, that rock mechanics is steadily changing corresponding to the evolution of experimental methods and of theory.

Table 1.1: Variables used in this paper

$\sigma_1, \sigma_2, \sigma_3$	principal stresses
σ_1, σ_z	axial stress
$\sigma = 1/3(\sigma_1 + \sigma_2 + \sigma_3)$	mean stress, octahedral normal stress
$s_i = \sigma_i - \sigma$	stress deviator
$\tau = (1/3\,(s_1{}^2 + s_2{}^2 + s_3{}^2))^{\frac{1}{2}}$	octahedral shear stress
$\quad = 1/3\,((\sigma_1 - \sigma_2)^2 + (\sigma_2 - \sigma_3)^2 + (\sigma_3 - \sigma_1)^2)^{\frac{1}{2}}$	
$\Delta\sigma = \sigma_i - p$	eqivalent stress, differential stress, or stress
$\quad = 3/\sqrt{2}\cdot\tau = 2.12\cdot\tau$	difference in a Karman test
τ_B	failure strength
τ_R	residual strength
$m = \dfrac{3\,s_2}{s_1 - s_3}\ ;\quad s_1 > s_2 > s_3$	stress path, load geometry, Lode parameter
$J_m = \dfrac{m\cdot(9 - m^2)}{(3 + m^2)^{\frac{3}{2}}}$	invariant of stress geometry
$\varepsilon_i = (l_0 - l_i)/l_0$	principal technical strains
$\varepsilon v = (V_0 - V)/V_0$	relative volume change
l_0	initial length
l, l_i	actual length
V_0	initial (reference) Volume
V	actual volume
p	confining pressure in triaxial Karman tests
$\dot{\varepsilon}$	axial strain (or deformation) rate
$\dot{\varepsilon}_s$	steady state creep rate
$\varepsilon = \ln(l/l_0)$	true strain
$\dot{\sigma}, \dot{\tau}$	stress rates
$\theta = \partial\sigma/\partial\varepsilon$	strain hardening rate
T	temperature
F	applied (axial) force
A_0	initial cross section
A	actual cross section
A_g	cross section of vessel opening for the piston
W^I	irreversible stress work
d	average subgrain diameter

2. QUASISTATIC UNIAXIAL, TRIAXIAL, AND TRUE TRIAXIAL LABORATORY TESTS

2.1 General remarks

Before designing tests and equipment for testing one has to consider, what kinds of experiment have to be carried out and what the purpose of the tests is. In our case we deal with geomaterials, i.e. rocks and similar materials, and their use in engineering geology. The final aim is, in general, the design of constitutive laws and the definition of appropriate parameters for these materials which can be used in model calculations, e.g. finite element models.

Thus the question arises: what is rock?

On a small scale it consists in most cases of crystal grains with grain sizes in the range of millimeters and centimeters. They are jointed together with or without a cement. The grains can consist of different kinds of minerals with different mechanical behaviour. On a larger scale cracks, joints, bedding planes, or different strata appear. The result is that rocks do not behave homogeneously as metals usually do. Under certain conditions this results in a rather complicated mechanical behaviour because stress and strain distribution tend to be unequal on a small as well as on a large scale, often causing fissuring and loosening. When carrying out tests we have to take this into consideration. This must be regarded for instance when we choose the sample size. Often a minimum number of 2000 grains is regarded to be sufficient for one specimen. Therefore, a diameter of 10 cm and lengths of 20 or 25 cm are quite common for rocks. One has also to consider what the general aim of the tests is: a certain engineering aim or perhaps a basic research question.

Let us first see in table 2.1 what kind of laboratory tests are common:

Table 2.1: Table of common laboratory tests on rocks (continued on next page)

quasistatic tests: duration a minute to an hour,
    stress rate or strain rate control, uniaxial or triaxial,
    usually failure tests
creep tests: duration hours to years,
    constant stress, uniaxial or triaxial
relaxation tests:
    stop of deformation after some strain, uniaxial
dynamic tests: milliseconds to minutes,
    high strain rates or cyclic tests, mostly uniaxial
Tests for determining the elastic constants:
    Unloading tests, elastic wave velocities
tension tests:
    uniaxial or triaxial

Table 2.1 (continued)

special tests:
    Brazilian test (indirect tension test)
    3-point or 4-point bending test
    fracture toughness test, fracture resistance test
    torsion test
    shear test
    biaxial test
    stress drop test
    hydrostatic compaction test at low or high hydrostatic pressure
special measurements:
    volume change
    permeability
    microacoustics (AE: noise related to microcracks)
    wave velocity and attenuation
    electrical resistivity
    pore pressure influence
    moisture influence
    influence of end friction and of specimen shape or size

This chapter deals with quasi-static tests, the necessary equipment and the experimental procedures. Quasi-static tests are tests which are slow enough not to cause dynamic effects. These experiments are mainly carried out to determine failure strength (ultimate stress) under stress rate control or strain rate control. However they can be used for many more objectives. It has to be stressed that one should in general use only true strain rate $\dot{\varepsilon}$ and true strain ε. This has been done throughout this publication.

For clarification it is mentioned that "compression test" means a test where the axial stress σ_1 is larger than the confining pressure p (specimen shortens) whereas "extension test" means a test where the axial stress σ_1 is smaller than the confining pressure p (specimen elongates).

2.2 Apparatus and test procedures for uniaxial and triaxial quasistatic tests

Let us start with uniaxial tests. These tests are mostly carried out with stress or load control with, say, $\dot{\sigma} = 1$ MPa/min rate of stress increase. A test rig for this kind of tests is not shown since its design is rather simple. Figures are shown for triaxial rigs. In uniaxial tests the sample is loaded by a piston and the oil pressure in a cylinder which usually is controlled by a servocontrol. Screw-driven tests rigs are also rather common. Often displacement is measured by three LVDT-transducers (transducers working on the basis of a variable transformer for electrical tension) separated by angles of 120 degrees around the specimen and parallel to its axis; strain gages are not recommended for soft geomaterials like rock salt. Experience shows that uniaxial quasistatic tests on rocks can yield results with large scatter caused by the inhomogeneity of the material. In order to remove most of the scatter it is recommended to reconsolidate the specimen before the test for removing the damage caused by drilling and sample preparation. This is very efficient with rock salt but does not work well with hard rock.

In general, triaxial tests should be preferred for failure tests because scatter is much reduced, because the natural state of stress is a triaxial one, and because one receives a much more complete picture of the mechanical behaviour of the rock. But the experimental technique is more sophisticated than for uniaxial tests. In most cases Karman-type rigs are used; very many kinds of this type have been constructed since the beginning of the century. Figures 2.1 and 2.2 show examples. The rig in figure 2.1 consists of a pressure vessel for the confining pressure p, a cylinder with a piston for the axial force F or the axial stress σ_1, three LVDT-transducers for deformation measurement mounted between the platens near to the sample, and heaters inside and outside of the pressure vessel. In this example the vessel is placed in a frame which supports the axial load. This is not needed for a self-

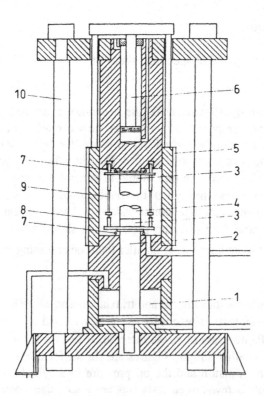

Figure 2.1: Triaxial testing machine for cylindrical specimens used at BGR.
1: hydraulic cylinder, 2: loading piston, 3: platen, 4: specimen, 5: external heater, 6: lift cylinder for vessel, 7: internal heater, 8: pressure vessel, 9: deformation transducer (LVDT), 10: rod.

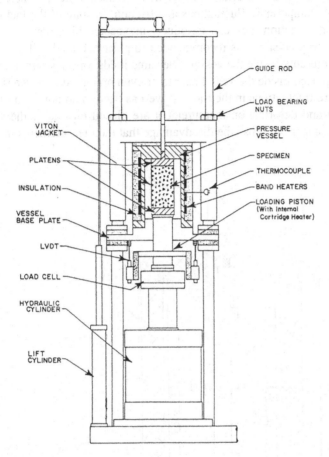

GUIDE ROD

LOAD BEARING
NUTS

VITON
JACKET

PRESSURE
VESSEL

PLATENS

SPECIMEN

THERMOCOUPLE

INSULATION

BAND HEATERS

LOADING PISTON
(With Internal
Cortridge Heater)

VESSEL
BASE PLATE

LVDT

LOAD CELL

HYDRAULIC
CYLINDER

LIFT
CYLINDER

Figure 2.2: Triaxial testing machine used at Re/Spec and Sandia (USA).

supporting vessel. This rig is not equipped with a load cell for measuring the axial load, but this is the case in figure 2.2 which has a somewhat different appearance but more or less similar characteristics. Experience shows that, due to the friction of the piston seal, the determination of the axial load from the oil pressure applied to the piston is less precise than the measurement by a load cell. In addition there are two servocontrol systems for confining pressure and axial load, a temperature control, and a data aquisition system.

These two apparatus have been constructed mainly for tests on rock salt. The characteristics of the testing machine in figure 2.1 are: T = 20 to 80°C, p = 0 to 45 MPa, F = 0 to 400 kN, standard sample size: 10 cm diameter, 25 cm in length. The characteristics of the testing machine in figure 2.2 are: T = 20 to 200°C, p = 0 to 70 MPa, F = 0 to 1500 kN, standard sample size: 10 cm diameter, 20 cm in length. Maximum displacement is about 80 mm in both cases. The rigs can also be utilized for creep tests using control systems designed for long therm tests (see chapter 5).

Some remarks have to be made concerning the accuracy of such systems. There exist two "enemies of precision": temperature fluctuations and friction. In spite of the fact that modern seals have quite low friction they can introduce unexpected high errors on the axial load. Therefore, for high precision it is recommended to place the load cell as near to the sample as possible, preferably inside the vessel. The same holds for the strain measurement which should also be placed inside the vessel near to or even on the sample (see figure 2.1). In this case temperature fluctuations in the room as well as friction do not distort the measurements. If load cell and deformation measurement are placed outside of the vessel the construction may be cheaper but it has the disadvantage that the measurements may be less

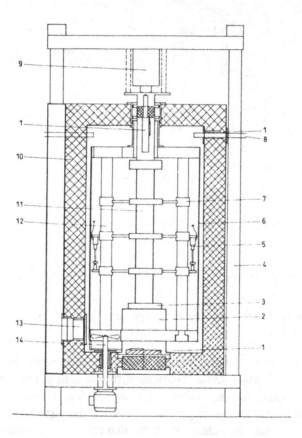

Figure 2.3: Uniaxial testing machine for four simultaneous creep tests with an external chamber for heating (22 to 80°C) used at BGR. Air humidity (0 to 75%RH) can be controlled as well.
1: heater, 2: hydraulic cylinder, 3: platen, 4: external rod, 5: LVDT, 6: band heater, 7: guide with ball bearings, 8: supply for humid or dry air, 9: load cell, 10: chamber with insulation, 11: specimen, 12: guide rods, 13: opening for electric wires, 14: fan.

precise. To avoid the influence of temperature fluctuations and temperature differences on the results the rig sometimes is placed in a heated chamber. This is shown for a uniaxial testing machine in figure 2.3 which can be used for temperatures up to 80°C and which allows control of air humidity at the same time. Anyway, it cannot be stressed often enough that careful calibration is the heart of reliable measurements.

The usual way of carrying out quasistatic tests in a Karman cell is as follows: an initial small hydrostatic stress is applied to the sample by a confining pressure p and an appropriate axial load σ_1. Then the sample is heated to the desired value. Finally, the test itself is performed by increasing the hydrostatic pressure to the desired value; then the axial load is raised with stress or strain control, if one wants to perform a compression test. An extension test is performed by decreasing the axial load. The specimen is elongated in such a test since the confining pressure is greater than the axial stress. It has to be taken into account in triaxial tests that the confining pressure reduces the axial load, being a function of the cross-section of the sample and of the vessel opening for the piston. If one does not use an internal load cell this must be taken into account using the following formula:

$$\sigma_1 = \frac{1}{A} \cdot \left[F + p \cdot \left(A - A_g \right) \right] \qquad \text{effective axial stress} \qquad (2.1)$$

$$A = A_0 \cdot \frac{l_0}{l} \qquad\qquad\qquad \text{present cross section of specimen}$$

$$A_g \qquad\qquad\qquad\qquad \text{cross section of vessel opening for the piston}$$

The other variables are given in table 1.1.

Two assumptions are involved in this formula: (1) no volume change occurs in the specimen, and (2) the specimen keeps its cylindrical shape. However, some rigs have a hydraulic compensation for the effect of p on σ_1 (see figure 3.11). Some rigs for uniaxial tests have a mechanical compensation of the effect of the change of the cross-section of the specimen on σ_1 by a specially curved cantilever as shown in figure 2.4 (Blum & Pschenitzka 1976, Vogler 1992). This device adjusts the axial force according to the deformation of the specimen and keeps a constant σ_1 during the test.

As an example, figure 2.5 shows test results for the influence of the confining pressure p on stress-strain curves of rock salt for a certain deformation rate $\dot{\varepsilon}$. Strength and deformation at failure increase considerably with increasing p for a soft kind of rock. Figure 2.6 shows the result of a test series carried out on marble, a more brittle kind of rock. Again, strength is increasing with increasing p; however, deformation at failure is only slightly dependent. It must be mentioned that these tests are often combined with the measurement of the elastic constants by unloading and reloading (see chapter 4.1) or with one of the special measurements mentioned in table 2.1.

An important topic is the specimen preparation. It is essential to prepare the samples with correct dimensions, in particular for triaxial tests. The surfaces must be at right angles

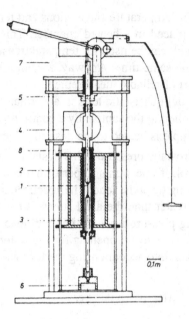

Figure 2.4: Rig for uniaxial creep tests with a specially curved cantilever for keeping constant σ_1 during a test, here for tension (Blum & Pschenitzka 1976).

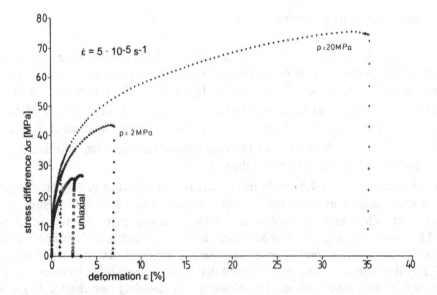

Figure 2.5: Results of uniaxial and triaxial compression tests on rock salt from the Asse mine (Northern Germany) at the same deformation rate but different confining pressures (BGR).

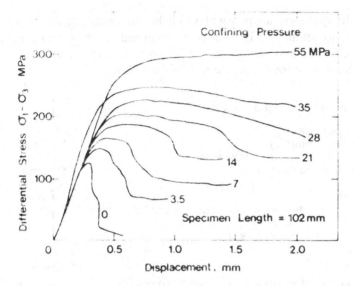

Figure 2.6: Results of uniaxial and triaxial compression tests at the same displacement rate $v = 10^{-3}$ mm s^{-1} ($\dot{\varepsilon} = 1 \cdot 10^{-5}$ s^{-1}) but different confining pressures for Tennesse marble (Paterson 1978, p. 144).

with each other and must be plane. A precision of 0.01 mm for the end faces is possible using a lathe. The rubber or metal jacket used for protection of the specimen against the hydraulic oil in a triaxial test must be fixed carefully, especially for extension tests. Usually the stiffness of this material is negligible and does not influence the stress in the specimen. It is also important to document the state of the specimens carefully before and after the tests. This makes it possible to carry out statistical analyses e.g. of the influence of grain size, petrography, or texture. Sample storage may also be delicate. For instance, change of moisture content may change the mechanical properties considerably. An effective protection against change of moisture content is only possible by a metal cover.

2.3 True triaxial testing equipment and test procedure

There exists another more advanced technique for triaxial tests: the so-called true triaxial tests. They can be done using thin-walled hollow cylinders or using cubic specimens. In the following, an equipment for cubes is described. Tests of this kind have the great advantage that all kinds of stress geometries can be simulated which occur in mother nature. As we will see later this is of considerable importance for rocks.

The true triaxial test rig of the BGR consists of a rigid frame in which six double-acting pistons are arranged opposite each other about the center of the frame (figure 2.7). To guarantee balanced forces on each axis the two pistons of each axis are hydraulically connected. The weight of the pistons for the vertical axis is equalized by counter weights and a hydraulic counter balance. The force applied on each axis can be controlled separately. A load is applied to the cubic specimens (up to 20 cm to a side) by six pistons via square steel platens. The specimens are prepared on a lathe. Six independent PID regulators control the temperature to within 1K. The forces are regulated via three pressure gauges. Three independent electronic units control load and deformation via three servovalves. The most important specifications of the rig and the tests are as follows:

Table 2.2: Specifications of the true triaxial tests

```
- Maximum force (per axis):                        2000  kN
- Sample size (edge length):                         53  mm
                         maximum:              ca.  200  mm
- Platen size in standard tests:                     50  mm
- Temperature            maximum:                   400  °C
- Friction of piston:                        < 1 % of load
- Independent load or deformation control along the 3 axes is
  possible.
- Loading rates in standard tests:
  hydrostatic phase:                     σ̇ =  7.6 MPa/min
  deviatoric phase                       τ̇ = 21.4 MPa/min
```

- Deformation rate at failure, standard tests: $\dot{\varepsilon}_{eff} \approx 0.007 \text{ s}^{-1}$

```
- Digital data acquisition, standard interval:        1  s
- Lubrication with a thin film of paraffin wax (for tests at
  room temperature) or graphite (elevated temperatures)
```

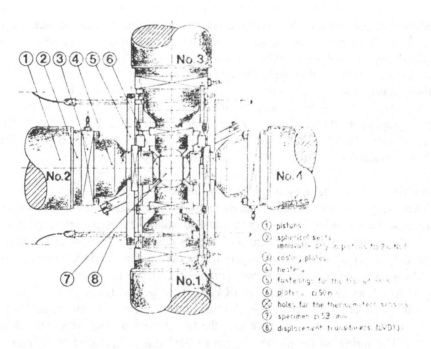

Figure 2.7: Arrangement of specimen and pistons in the true triaxial press for cubes at BGR (Hunsche & Albrecht, 1990).

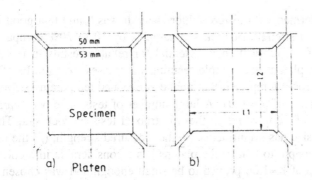

Figure 2.8: Cross section of the arrangement of specimen and platens in the true triaxial press of BGR (to scale); a) at beginning; b) at failure (for rock salt).

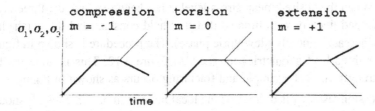

Figure 2.9: Schematic of the test procedure in true triaxial tests on cubes for three load geometries.

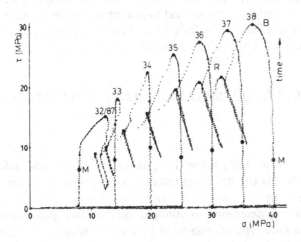

Figure 2.10: Seven true triaxial compression tests of test series S8 on rock salt from the Gorleben salt dome (Northern Germany) performed at room temperature at BGR. Spacing of points: 1s. M: minimum volume; B: failure; R: residual strength.

Many tests have been performed with different lubricants. It was found that good lubrication decreases the measured strength of rock salt samples by 10 to 20% compared with non-lubricated ones; therefore, lubrication is essential for reliable results in this kind of tests since friction between platens and sample seemingly increases strength. Paraffin wax is an effective lubricant at room temperature, graphite at elevated temperatures. Both have very small friction coefficients of $\mu \approx 0.01$. A large number of tests have been carried out to determine the influence of sample size in the true triaxial tests on rock salt. The main results are: (1) the sample size has no influence on the measured strength; (2) the ratio (K) of the edge length of the sample to that of the platens has a considerable influence on the measured strength. A value of K=1.06 proved to be small enough and was chosen for the tests. Figure 2.8 shows a cross section of the arrangement of specimen and platens. A precise calibration makes it possible to determine the volume change of the sample during a test with an accuracy of 0.3%.

It is advisable to carry out the true triaxial tests in a deviatoric manner. The following procedure is therefore recommended. In the tests, the load is first applied hydrostatically (hydrostatic phase). When the desired mean stress level σ has been reached, the three principal stresses are changed linearly with time so that σ is held constant and the octahedral shear stress τ is also increased linearly (deviatoric phase). The procedure is shown in figure 2.9 schematically for three load geometries m; see also figure 3.12. This is done until τ reaches the failure strength τ_B of the sample and fracturing occurs as shown in Figure 2.10 for one set of compression tests. τ_B is defined as the local maximum of τ in a test. It should be noticed that the shear stress does not go to zero at failure; the fractured sample has a considerable residual strength τ_R which can be determined by unloading and reloading as shown in figure 2.10 (Hunsche 1992).

Thus, some important tools for the performance of uniaxial and triaxial tests on rocks have been shown. Analysis and results are shown in chapter 3. It has to be emphasized that much more kinds of rigs have been constructed and that more testing procedures are possible. Anyway, it is important to concider carefully what kinds of tests are needed and with which test procedure the results under question can be derived.

3. ANALYSIS AND RESULTS OF QUASISTATIC FAILURE AND DEFORMATION TESTS

3.1 Tests without tension

In the previous chapter it has been shown how quasistatic failure and deformation tests should be carried out. In this chapter some results will be given. However, it is not possible to give a complete overview since there exists a huge amount of literature. The final aim of the experiments is to determine constitutive equations and parameters for the mechanical behaviour of the rock type of interest to be used in model calculations. One element of these equations is the determination of yield surfaces.

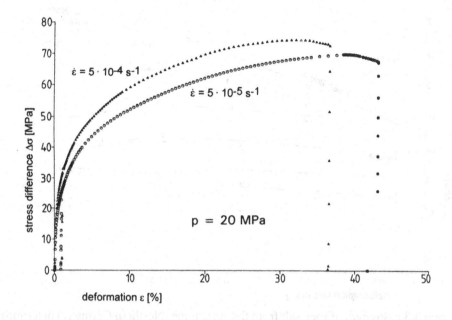

Figure 3.1: Triaxial compression tests on rock salt from the Asse mine (Northern Germany) at the same confining pressure p but at different deformation rates $\dot{\varepsilon}$ (BGR).

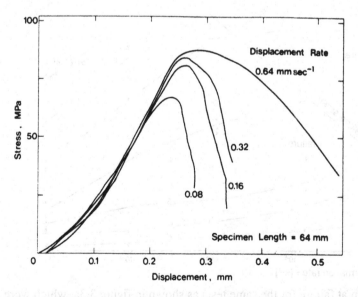

Figure 3.2: Uniaxial compression tests on Arkose Sandstone at different deformation rates $\dot{\varepsilon}$ from $1.25 \cdot 10^{-3}$ to $1 \cdot 10^{-2}$ s^{-1} (Paterson 1978, p.145).

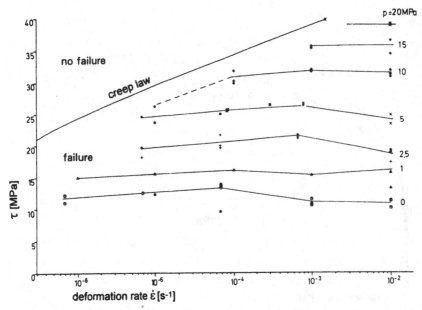

Figure 3.3a: Strength of rock salt from the Asse mine (Northern Germany) determined from triaxial compression tests on cylindrical specimens at a great number of confining pressures p and deformation rates $\dot\varepsilon$ at room temperature (BGR). The line denoted with "creep law" represents the $\dot\varepsilon$ - τ conditions for steady state creep.

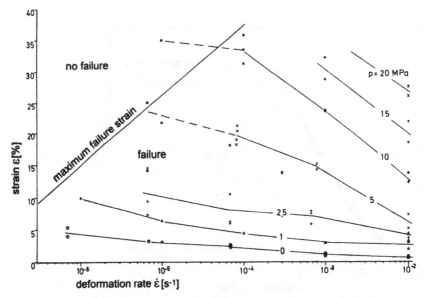

Figure 3.3b: Strain at failure for the same tests as shown in figure 3.3a, which were performed on virgin samples. There was practically no failure above the line denoted with "maximum failure deformation".

In most cases, quasistatic experiments are carried out with deformation control where $\dot{\varepsilon}$ is held constant. At a fixed confining pressure p but for various $\dot{\varepsilon}$ one receives $\Delta\sigma$ - ε - curves as shown in figure 3.1. For rock salt, failure strength (ultimate stress) as given by the peak point, is not much dependent on $\dot{\varepsilon}$. This is somewhat different for more brittle types of rock where strength increases considerably with increasing $\dot{\varepsilon}$ as shown in figure 3.2 (Paterson 1978, p. 145). However, due to the ductile behaviour of rock salt the deformation at failure is much more dependent on $\dot{\varepsilon}$ than for brittle rock. The following has to be pointed out: if a test is carried out at a sufficiently low value of $\dot{\varepsilon}$ than the experiment becomes a creep test without short term failure (but perhaps with tertiary creep after a longer period of time). This is shown in figure 3.3a in a more complete picture of the failure behaviour of rock salt derived from a great number of compression tests on one rock salt type with various values for constant $\dot{\varepsilon}$ and at various confining pressures p. In this diagram strength is given as a function of p and $\dot{\varepsilon}$ for room temperature (see also Wallner 1983). On the left side the failure domain is bounded by the steady state creep curve for 22°C. Sometimes this limit is named as brittle-ductile transition. But in other cases this transition is defined differently (see Paterson 1978). It is obvious from figure 3.3a that this type of rock salt has a weak maximum in strength at about $\dot{\varepsilon} = 10^{-4}$ s^{-1} for all confining pressures. This is different for hard rocks where strength increases. This result has also been derived by Klepazcko, Gary & Barberis (1991) who have performed uniaxial compression tests on rather impure rock salt between $\dot{\varepsilon} = 3{\cdot}10^{-5}$ s^{-1} and $4{\cdot}10^{3}$ s^{-1} as drawn in figure 3.4. In addition they find a minimum in strength at about 2 s^{-1}. This is shown to demonstrate that failure does not always follow the anticipated rules. In addition, figure 3.3b shows the strain at failure for the tests in figure 3.3a. Obviously, this strain is dependent not only on p but also on $\dot{\varepsilon}$ since the specimen has more time to deform in the slow tests than in the fast ones. These strains are

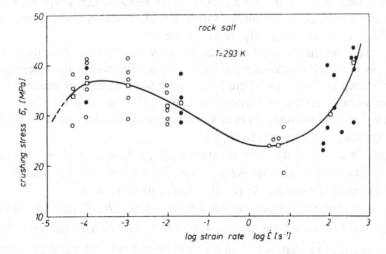

Figure 3.4: Failure strength of rock salt derived from uniaxial tests performed with strain rates between $\dot{\varepsilon} = 3{\cdot}10^{-5}$ s^{-1} and $4{\cdot}10^{3}$ s^{-1} (Klepaczko, Gary & Barberis 1991).

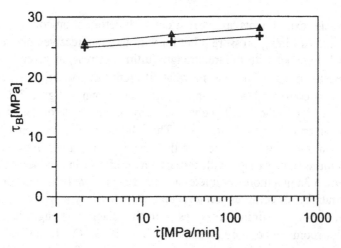

Figure 3.5: Strength of rock salt from the Gorleben salt dome (Northern Germany) determined in triaxial compression tests on cubic samples for three rates of stress change $\dot{\tau}$ at room temperatuie (BGR). Mean stress at failure was $\sigma = 25$ MPa.

different for different types of rock salt. The "maximum failure deformation" gives the limit for deformation at failure for certain confining pressures. Of course, this limit can be exceeded in creep tests.

If one carries out failure tests with constant rate of stress change $\dot{\sigma}$ or $\dot{\tau}$ one can determine the dependence of strength on this parameter. Figure 3.5 shows the results of such tests derived from compression tests on cubic rock salt specimens. With increasing $\dot{\tau}$ the strength τ_B is only slowly increasing. It has to be mentioned that in such kind of tests failure is finally associated with the maximum speed of the piston which again is caused by the maximum pumping rate of the hydraulic system. Therefore, it is not very surprising that the influence of $\dot{\sigma}$ or $\dot{\tau}$ on strength is only small in these tests.

From quasistatic deformation tests one can also draw conclusions about the creep behaviour and the microscopic mechanisms, since a considerable part of the deformation is due to (transient) creep in soft rocks. There have been made a number of attempts for this purpose, one has been described by Mecking & Estrin (1987) and by Hertzberg (1983). Their argument is that there are four stages of strain hardening in a deformation experiment performed with constant $\dot{\varepsilon}$ as shown in figure 3.6:

I: easy glide (only in single crystals); II: constant strain hardening rate $\theta = \partial\sigma/\partial\varepsilon$ during multiple slip in polycrystals without recovery (creep); III: decrease in θ because plastification (recovery and creep) is initiated; IV: possible stage with constant θ.

Stage III can be described by the equation: $\Delta\sigma = \sigma_s \cdot [1 - \exp(-\varepsilon/\varepsilon_s)]$, where σ_s and ε_s are appropriate constants. It follows from this equation: $\theta = \theta_0 \cdot (1 - \Delta\sigma/\sigma_s)$. In figures 3.7 and 3.8 this has been applied to a triaxial test on rock salt carried out with constant $\dot{\varepsilon}$ and p. It can be inferred that there exist two phases in stage III with perhaps two hardening or

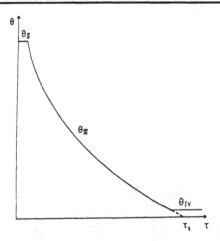

Figure 3.6: Schematic representation of strain hardening stages in a $\theta - \tau$ diagram (Mecking & Estrin 1987).

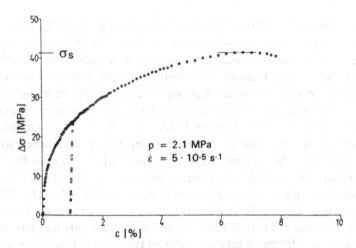

Figure 3.7: Triaxial failure test on rock salt from the Asse mine (Northern Germany) performed at constant deformation rate $\dot{\varepsilon}$ at room temperature (BGR).

recovery mechanisms in this test. This finding has to be further evaluated. However, the example indicates that even relatively simple and fast deformation experiments can yield insight into the deformation mechanisms.

After all, one has to keep in mind that most of the common tests on rocks are carried out in compression, i.e. the sample becomes shorter. But mother nature uses all kinds of stress geometries as defined by the Lode parameter m. Examples are compression (m=-1), extension (m=+1), or torsion (m =0). A complete picture of strength behaviour can only be determined using true triaxial tests e.g. on cubic samples which have already been de-

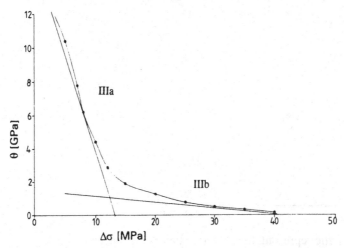

Figure 3.8: Strain hardening rate θ versus equivalent stress Δσ derived from the test in Figure 3.7.

scribed in chapter 2.3. A number of test series on various types of rock salt yield the results for failure strength shown in figure 3.9 as a function of mean stress σ for different values of m. The main results are: strength increases with increasing σ, as one also finds for other kinds of rock, and extensional strength is about 30% smaller than compressional strength. This difference is decreasing at higher σ which has been found for hard rocks as well. Moreover, there exist distinct differences in strength for different rock salt types of about 4 MPa or 30% (Hunsche & Albrecht 1990, Hunsche 1992, 1993). If one fits an empirical equation to the lower boundary of these results as shown in figure 3.9 (i.e. in a conservative way) one receives the conservative failure equation given in table 3.1. It describes the dependence of strength on mean stress σ, Lode parameter m, and temperature T. The effect of temperature becomes visible only above 100°C (Hunsche & Albrecht 1990, Hunsche 1992, 1993). This function is drawn in figures 3.10 and 3.9. It is ready to be used in model calculations. If one wants to fit the measured data directly this should be done only for tests on a single type of salt because of the differences between different types of salt (see figure 3.9). There are no measured true triaxial data below the lower limit of validity in figure 3.10 since tension has to be applied in this range in one or two directions for failure. Tensional tests will be regarded in capter 3.2 in combination with biaxial tests. But it is a conservative and safe assumption that the tensional strength is zero at low values of σ.

However, it cannot be the final aim to determine strength of rock only in terms of stress. A comprehensive picture should also include the prefailure behaviour and this does not only mean deformation but also volume change and its related phenomena: permeability, creep rupture, damage, microacoustic emissions, and others. Because this is especially substantial in repositories, the determination of dilatation is becoming an increasingly important field in rock mechanics. Dilatation has been observed in many kinds of rock and soil (e.g. Paterson 1978, Cristescu 1989, Cristescu 1994 this volume).

Table 3.1: Conservative failure equation for rock salt (Hunsche 1992. 1993)

$\tau_B = f(\sigma)\cdot g(m)\cdot h(T)$ <u>failure strength</u>

$g(m) = 2k/[(1+k) + (1-k)J_m]$

$h(T) = \begin{cases} 1 & \text{for } 20°C \le T \le 100°C \\ 1 - c(T - 100°C) & \text{for } 100°C \le T \le 260°C \end{cases}$

$b = 2.7$ MPa $p = 0.65$

$k = 0.74$ $\sigma_* = 1$ MPa

$c = 0.002$ K^{-1}

$\tau_R = \tau_B(m=+1)$ <u>residual strength</u>

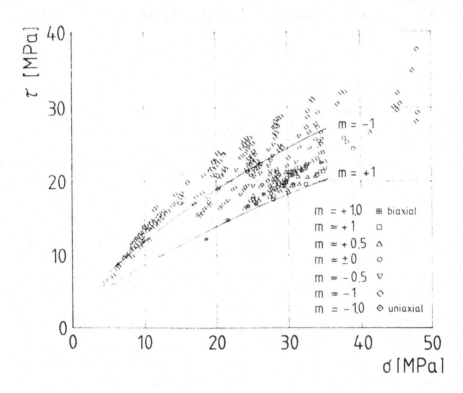

Figure 3.9: Failure strength for 12 series of true triaxial tests on different rock salt types from the Gorleben salt dome (Northern Germany), carried out at BGR, and grouped according to the load geometry m; T = 30°C. The lines represent the conservative failure equation in table 3.1 for m = -1 (compression) and m = +1 (extension) (Hunsche 1992, 1993).

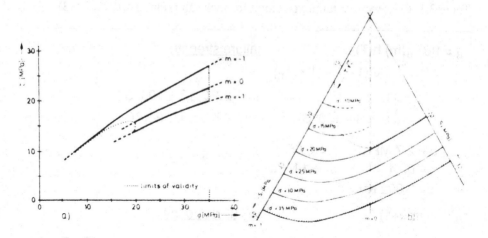

Figure 3.10: Plots of the conservative failure equation for rock salt (table 3.1) at T ≤ 100°C: (a) in the τ – σ plane; (b) in the octahedral plane (Hunsche 1992, 1993).

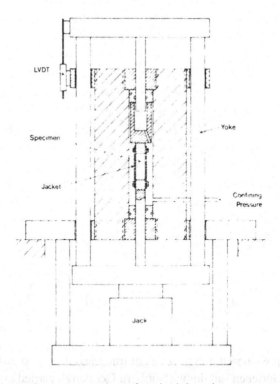

Figure 3.11: Schematic representation of a triaxial testing apparatus incorporating Grigg's arrangement of compensating piston (Paterson 1978, p. 8).

Volume change can be measured by means of different kinds of devices:

- It can determined from tests on cubic samples where disturbances at the edges are removed by careful calibration (see figure 2.8).

- Volume change can be measured by means of the expelled oil used for the confining pressure in a Karman type test on a cylindrical specimen. This is not a simple method because oil is also displaced by the moving piston for the axial load. A compensation for this amount of oil is advisable, therefore. One principle for a compensation is shown in figure 3.11. This is accomplished by moving the closed yoke as a whole, which has also the advantage that only the stress difference $\Delta\sigma$ has to be applied by the pistion. It has to be mentioned that temperature disturbances can introduce considerable errors in the measurement of volume change.

- Volume change can also be measured by means of lateral strain measurement devices directly fixed to the specimen. Possibilities are: strain gauges glued on the sample, a metal ring fixed around the sample and equipped with strain gauges, or a tight ring worn

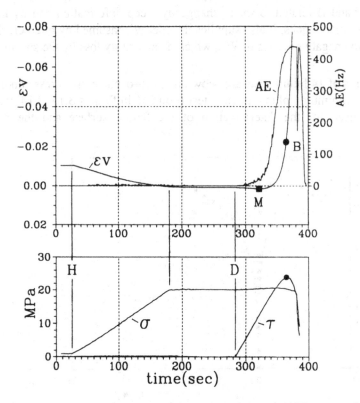

Figure 3.12: Volume change εv and acoustic emission rate (AE) versus time in a true triaxial compression test performed at BGR on rock salt from the Gorleben salt dome (Northern Germany). (Hunsche 1992, 1993).
M: minimum volume; B: failure;
H: begin of hydrostatic loading phase; D: begin of deviatoric loading phase.

around the sample equipped with a deformation transducer of any kind. Anyway, one should use more than one of these devices because lateral deformation tends to be inhomogeneous and the barrelling of the specimen has to be considered as well.

Depending on the stress state the volume of a specimen increases (dilatancy) or decreases (compressibility, contractancy) due to the overall microcrack opening or closing. In figure 3.12, the relative volume change εv is shown for one of the true triaxial tests on cubic samples, described in chapter 2.3, together with the acoustic emission rate (AE), which is obviously a good indicator for microcrack opening and volume increase. It should be noted that the determination of volume change in the true triaxial tests has been quite difficult to perform. Figure 3.13 shows strains ε_i and volume change εv of a cubic sample during the deviatoric phase of a test. In addition, results are drawn as calculated with the nonassociated elastic-viscoplastic constitutive equation given by Cristescu & Hunsche (1992, 1993a,b) and by Cristescu (1994, this volume). They fit the measurements very well. In figure 3.14 the irreversible stress work for a test is given (see Cristescu 1994 this volume, and Hunsche 1992). W^I_v is the energy related to volume change. W^I_d is caused by the stress deviator and is related to shape change by creep deformation and by initiation and propagation of microcracks; it also supplies the energy consumed by microcrack opening expressed by the negative values of W^I_d, which mean energy loss by the specimen. W^I_g is the sum of both.

In figure 3.15 results of 14 failure tests are shown in the $\tau-\sigma$ plane: the stress conditions at failure and for the minimum volume in each test (point M in figures 3.12, 3.14 and 2.10). They have been used for the determination of the failure surface and the dilatancy

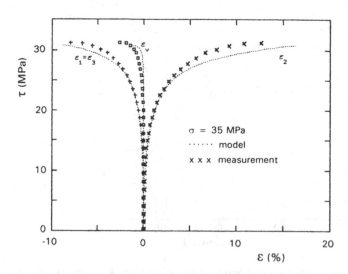

Figure 3.13: Strains ε_i and volume change εv of a cubic rock salt specimen during the deviatoric phase of a test. For test conditions see table 2.2.
In addition, the result of a model calculation is given. For details see Cristescu & Hunsche (1992, 1993a,b) and Cristescu (1994, this volume).

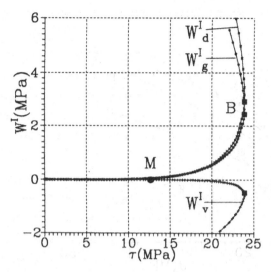

Figure 3.14: Irreversible stress work for a test on a cubic rock salt specimen during the deviatoric phase of a test (Hunsche 1992).
W^I_v: irreversible stress work related to volume change,
W^I_d: irreversible stress work related to shape change,
W^I_g: $W^I_v + W^I_d$

boundary, i.e. the boundary between the compressible and dilatant domain. These tests have been used by Cristescu & Hunsche (1992, 1993a,b) for the development of a comprehensive constitutive model for rock salt. A detailed description is given by Cristescu (1994, this volume). The shape of the dilatancy boundary is not yet clear for high mean stresses σ. Probably it extends more to the right. Moreover it may be more like a band in that range since the minimum volume is not very pronounced at high σ. At stress states above the dilatancy boundary, rock salt dilates and may increasingly loosen and fail after a period of time due to creep rupture. Failure occures when a certain amount of damage has been exceeded. To specify this condition, volume change or the irreversible volumetric stress work or even other variables can be used as a damage parameter. To give an impression af the amount of damage at failure, the volume increase at failure derived from the same tests is given in figure 3.16. Comparable values of volume increase at failure have been measured by Wawersik & Hannum (1980). An empirical equation for a damage parameter has been given by Cristescu & Hunsche (1993b) on the basis of the mechanical energy related to volume change. Below the dilatancy boundary, rock salt is compressed until all voids are closed. This is a time dependent process - volume creep - as shown in figure 3.17 for a purely hydrostatic compaction test ($\tau = 0$) with several hydrostatic loading steps of σ and a final unloading step. Under this stress condition the volume of the sample shrinks gradually during each loading step of more than one hour duration. Note also the slow volume increase in figure 3.17a after unloading and the rather nonlinear unloading curve in figure

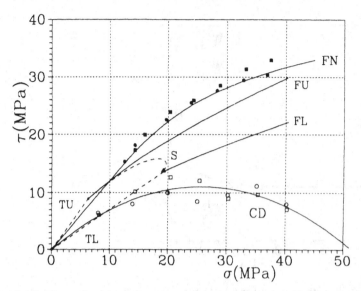

Figure 3.15: Plots of various equations and test results for rock salt from the Gorleben salt dome (Northern Germany) derived at 30°C, only for compression.

FN: failure equation (Cristescu 1993, 1994) and

● ■ : corresponding results from true triaxial tests (Hunsche 1992, 1993);

FU, FL: conservative failure equation for compression (m = -1) and
        extension (m = +1) in table 3.1 (Hunsche 1992, 1993);

CD: dilatancy boundary (Cristescu & Hunsche 1992, 1993a,b; Cristescu 1994) and
        ○ □ : corresponding test results (Hunsche 1992, 1993);

TU, TL: boundaries below which tension has to applied for failure;

S --- : lower limit of validity; ▲ : stress condition for biaxial compression tests.

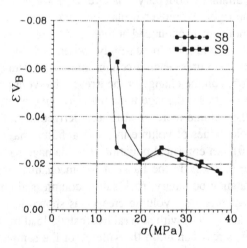

Figure 3.16: Volume increase at failure determined from the same true triaxial failure tests on rock salt as shown in figure 3.15 (Hunsche 1992, 1993).

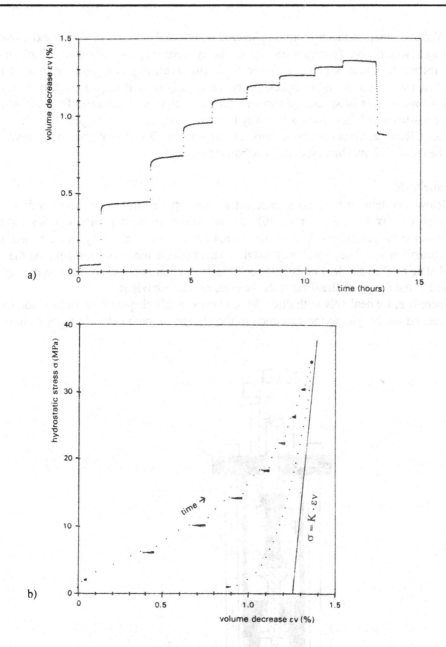

Figure 3.17: Result of a purely hydrostatic compaction test ($\tau = 0$) on a cubic rock salt specimen from the Gorleben salt dome (Northern Germany) with several hydrostatic loading steps of σ (BGR). Between the steps the stress was held constant for more than one hour.

a) volume change εv vs time; b) mean stress σ vs εv;

$\sigma = K \cdot \varepsilon v$: elastic volume compression.

3.17b. Volume creep is much more emphasized in crushed salt. There do not exist many publications which give formulas for the dilatancy boundary. A collection of dilatancy boundaries for rock salt is given in figure 5.12. However, this is a growing field of research, because dilatancy and compressibility are closely related to permeability of rock. Their determination is of special importance for the design of repositories for toxic or radioactive wastes which have to be kept away from the biosphere by tight geomaterials.

Post failure behaviour is not described in this course. But it is of great importance in certain fields of rock mechanics, e.g. in earthquake research.

3.2 Tensile tests

Below a certain mean stress σ tension has to be applied in one or two directions for failure (see figures 3.10, 3.15 and 3.20). In the vicinity of underground cavities failure occurs under these conditions. The tensile strength can be determined by indirect methods like the Brazilian test. They yield only rough results because their results depend on the individual state of the sample. For ductile rock the evaluation is not clear because the equations used in the theory of Brazilian tests presume elastic behaviour.

Therefore, we deal only with direct tension tests in this chapter. Uniaxial tension tests can be carried out by gluing the specimen to the platens on both sides. In order to have a

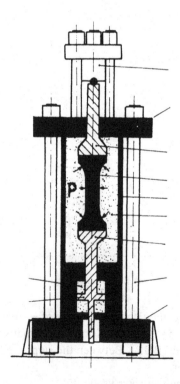

Figure 3.18: Schematic representation of a triaxial tension test performed on a dogbone-shaped specimen; length of sample: 25 cm (BGR).

suitable cross-section for the failure plane it is advisable to perform these tests on dogbone shaped samples as shown in figure 3.18 for a triaxial tension test. Uniaxial tensile strength of rock usually lies between 5% and 10% of the uniaxial compressional strength. For rock salt one receives about 2 MPa; the uniaxial compressive strength of rock salt is about 25 MPa. By using an idea of Brace (1964) tension tests can also be carried out under triaxial stress conditions on dogbone shaped samples. Thus it is possible to produce tensional stress along the axis of a cylindrical specimen in an ordinary Karman cell without the need to glue the sample at the ends, since the oil pressure in the vessel produces an axial tension at the flanks of the specimen as indicated in figure 3.18. The axial tension is then controlled in practice by the (compressive) axial load. The experimental technique is rather difficult because small differences in stress along the sample axis have to be determined. However, such experiments provide data over a wide range of confining pressures by the use of various sample shapes. At higher confining pressures p the triaxial tests are bounded by the biaxial tests which can easily be carried out on cubic samples. Figure 3.19 gives a set of results of uniaxial and triaxial tension tests and biaxial tests on one type of rock salt (Hunsche 1993). It is obvious that the (negative) axial tensile stress σ_z at failure is increasing with increasing p and becomes zero at p = 40 MPa for this type of rock salt. This is in good agreement with the results of biaxial tests shown. It even becomes positive in extensional failure tests. This finding has also been reported by Paterson (1978, p. 19). The reason for this surprising result is probably that microcracks and subsequent fractures origi-

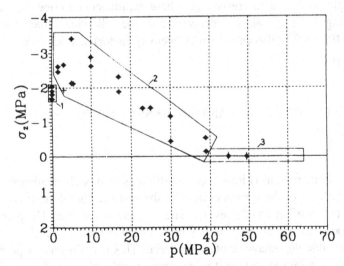

Figure 3.19: Plot of tensional failure strength σ_z as a function of p determined in uniaxial and triaxial tension tests on dogbone-shaped specimens, as well as results of biaxial tests on cubic specimens. σ_z is the stress along the sample axis. All tests were carried out at BGR at room temperature on one type of rock salt from the Asse mine (Hunsche 1993).
(1) uniaxial tension tests, (2) triaxial tension tests, (3) biaxial tests.

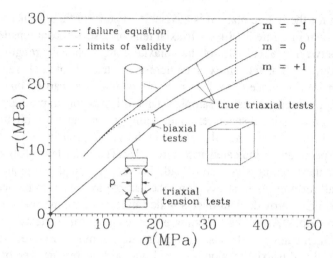

Figure 3.20: Different types of failure tests described in chapters 2 and 3.

nate in response to local tensile stresses around inhomogeneities like flaws or cracks. Deformation is practically zero in the tension tests. In figure 3.20, an overview is given over the tests discussed in chapters 2 and 3.

It has to be added that there exist numerous special experiments which are designed to measure fracture toughness or fracture resistance. These parameters describe the energy related to fracture opening. For rock salt one receives about the following values: fracture toughness $K_{IC} = 0.75$ MPa $\cdot$ m$^{0.5}$ (also called stress intensity factor) or fracture resistance $R = 1.5$ N $\cdot$ m^{-1} ($\approx [K_{IC}]^2/E$).

4. MEASUREMENT OF ELASTIC CONSTANTS, STRESS DROP TESTS, RELAXATION TESTS

4.1 Elastic properties

Knowledge of the elastic moduli is important for all kinds of model calculations, because it determines the stiffness of the system or parts of the system. Therefore, it is essential for the prediction of the distribution of stress and strain in space and time. The practical measurement of the moduli is not as easy as it seems, because the test results can be strongly influenced by the ductile behaviour of the material. This is not a serious problem for hard rock (if it does not contain joints) but it is one for soft rocks like salt. This problem can be overcome by performing fast experiments, were creep has no chance to come into effect. Therefore, the measurement of the elastic properties by means of ultrasonic waves yield the most logic results, from a physical point of view. For rock salt one receives the following values at room temperature:

$E = 36.7$ GPa, $G = 14.6$ GPa, $K = 25.7$ GPa, $\nu = 0.26$.

It has to be recalled that these values are temperature dependent. Frost & Ashby (1982) for example give the following formula for the shear modulus:

$$G = 15 \cdot [1 - 0.73 \cdot (T-300)/1070] \text{ GPa} ; \qquad T \text{ in K.} \qquad (4.1)$$

The elastic moduli can also depend to a certain extent on damage, i.e. the amount of micro-cracks in the rock (see chapter 3.)

In most cases Young's modulus E is measured by loading and unloading experiments. The test procedure is shown schematically in figure 4.1. Unloading tests yield a

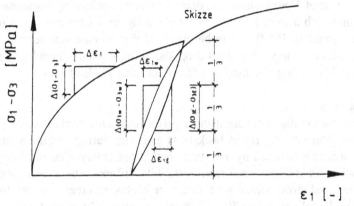

Figure 4.1: Schematic diagram showing the determination of the elastic modulus E from initial loading, unloading, and reloading.

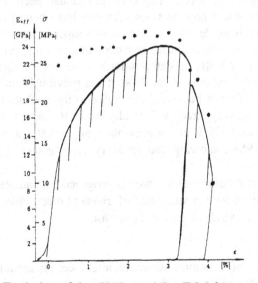

Figure 4.2: Evolution of the elastic modulus E [•] for a rock salt specimen with increasing strain determined from unloading cycles; $\dot{\varepsilon} = 10^{-4}$ s^{-1}, (Heemann 1989).

rather good estimation of the real value of E, if one uses the middle part of the unloading curve; for rock salt one receives values around 30 GPa by this means. In the upper part of the curve one observes still some foreward creep and in the lower part one observes some backward creep which is caused by the internal long range backstresses (see chapter 4.2). The determination of the E-modulus from loading tests on a virgin sample yields much smaller values since primary transient creep affects deformation strongly. Such a modulus should be named differently, "loading modulus" may be a better term. Values for the loading modulus between about 20% and 50% of the dynamic value for E are measured for rock salt. Nevertheless, these reduced values are often used in model calculations as a substitute for transient creep if this is not implemented.

The evolution of Young's modulus determined from unloading cycles as a function of deformation (and such also of damage) is shown in figure 4.2 for a uniaxial test on a rock salt specimen (Heemann 1989). It can be observed that an essential decrease of E occurs only in the post failure range; the stress drop should have been somewhat greater in these tests, however. This finding has also been found for other kinds of rock.

4.2 Stress drop tests

Stress drop tests (stress dip tests) are usually performed in materials sciences for the determination of the internal long range backstress evolving during creep. In such a test the axial stress is suddenly reduced by a certain amount preferably after reaching steady state creep. The backstress is ascertained from tests with different stress reductions and is defined as that reduced stress where no forward or backward creep occures for some time after stress drop. Usual values are 50 to 70% of the previously applied stress. Because the internal structure is not changed during and right after the stress drop they can also be used for the determination of internal variables (e.g. activation area). Transient creep laws can be formulated using the evolution of backstress. This concept has also been used for rock salt (see chapter 5). A number of stress drop tests on rock salt have been performed by Hunsche (1988) with the high-precision device for creep tests shown in figure 5.1. Figure 4.3 displays the results of different amounts of stress reduction after a long period of creep. After a stress drop of about 30% practically no deformation occures for a longer period of time. Therefore, the long range backstress is about 70% of the previously applied stress. After a large stress drop one can observe backward creep due to the backstress, even for several days and weeks. Due to recovery, creep will finally return to forward creep, however. In figure 4.4 stress drop tests to 70% of the previously applied stress is shown for several temperatures (Hunsche 1988), showing that recovery of long term backstress is temperature dependent.

Anyway, as the creep behaviour after a stress drop is governed by transient creep the result of these tests can also be used for the validation of transient creep laws in order to check the ability to describe extreme kinds of transient response.

4.3 Relaxation tests

Relaxation tests are uniaxial deformation tests where, after a certain amount of deformation, the relative movement of the endfaces of the specimen is suddenly stopped $(\dot{\varepsilon} = 0)$.

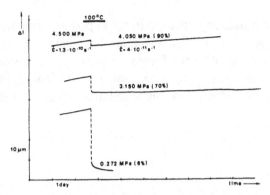

Figure 4.3: Uniaxial creep experiment on rock salt from the Asse mine (Northern Germany) with different stress drops (BGR). Sample length: 250 mm, measuring segment: 170 mm, (Hunsche 1988).

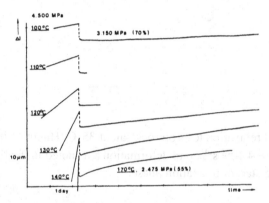

Figure 4.4: Uniaxial creep experiment on a rock salt specimen from the Asse mine (Northern Germany) with stress drops of 30% at different temperatures (BGR). Sample length: 250 mm, measuring segment: 170 mm, (Hunsche 1988).

Then the time dependent relaxation of the axial stress is measured. The relaxation is driven by the stored elastic energy and is controlled by the internal plastic deformation (transient creep under decreasing stress). Therefore, the lateral deformation is not zero during relaxation.

Relaxation tests are sometimes used in rock mechanics for the characterization of rock behaviour since stress reduction (and subsequent relaxation) also occures in the vicinity of underground openings. Relaxation can be described by the concept of rheological bodies. Relaxation tests can also be used to determine a lower creep limit; results of such tests and of other observations indicate that a creep limit does not exist for most materials. As in stress drop tests these tests have the advantage that the microscopic substructure is rather constant during the relaxation phase because practically no deformation induced recovery takes place.

It is important to state that relaxation tests are difficult to perform since the condition $\dot\varepsilon = 0$ is very delicate to be realized; Haupt (1988) mentions that a precision of 10^{-5} seems to be sufficient for the control of the axial length. Figure 4.5 shows the results of a number of relaxation tests carried out by Haupt (1988) on rock salt at 35°C. The preloading was performed at $\dot\varepsilon = 4\cdot10^{-5}$ s^{-1} up to the deformation ε indicated in the figure. In figure 4.6 the

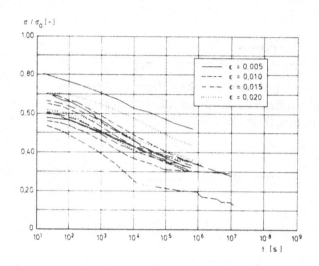

Figure 4.5: Results of uniaxial relaxation tests on rock salt at 35°C (Haupt 1988). Previous deformation rates: $\dot\varepsilon = 4\cdot10^{-5}$ s^{-1} up to deformation ε as noted in the figure. σ_a is the axial stress σ_1 before start of relaxation phase.

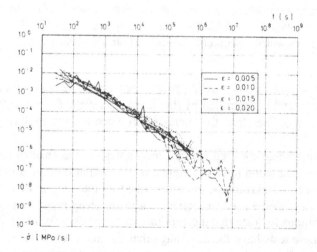

Figure 4.6: Results of the uniaxial relaxation tests on rock salt in figure 4.5 shown in another kind of diagram (Haupt 1988).

same tests as in figure 4.5 are shown in another kind of diagram: $\dot{\sigma}=f(t)$. One can observe that the slopes of all curves are similar and also that relaxation does not cease even after 10^7 s =121 d. This corresponds to a stress reduction down to less than 15% of the stress before relaxation σ_a and is equivalent to less than σ_1 =1 MPa in this special experiment. This means that a creep limit has not been found in these tests for rock salt. This can also be inferred from natural stress-strain conditions during the growth of salt domes (Hunsche 1978).

Since the creep behaviour after a relaxation test is governed by transient creep as in stress drop tests, one can use their results also for the validation of transient creep laws in order to check the ability to describe extreme kinds of transient response.

5. UNIAXIAL AND TRIAXIAL CREEP TESTS: EXPERIMENTAL TECHNIQUE AND RESULTS

5.1 Experimental technique

Knowledge of creep behaviour is a very important condition for reliable model calculations especially in soft rocks like rock salt. It is that part of geomechanical behaviour which influences most the time dependent strain and stress evolution in a system. Therefore, it is very important to determine the creep behaviour with sufficient accuracy. In the following, we refer mainly to rock salt as an example.

For uniaxial or triaxial creep tests the same apparatus are generally used as for quasi-static tests described in chapter 2. For creep tests some additional technical conditions must be regarded, however. Because creep tests on geomaterials usually have a very long duration of several mounths or even years all parts of the system must be stable during this long period of time. Special attention must be given to the following items. LVDT-transducers or optical systems have proved to be long term stable for deformation measurement whereas strain gages directly glud on the sample may be unstable due to creep effects in the glue. Pressure gauges and load cells must be of high quality, their drift and temperature effects must be minimized. Some thermometers can exhibit drifts as well. Therefore, resistance thermometers (e.g. PT 100) are recommended, thermocouples are not free of drift. Also electronic signal conditioners can exhibit considerable drift and have to be of high quality. The temperature influence on all mechanical and electronical parts of the system must be minimized since fluctuations of room temperature are nearly unavoidable during long term experiments. Periodical calibration of all parts of the system is essential.

It is of particular importance to control the stresses in the specimen with high accuracy during the tests, since the influence of the equivalent stress $\Delta\sigma = \sigma_1 - p$ on strain rate is very nonlinear. A stress exponent of about n = 5 {i.e. $\dot{\varepsilon}_s \sim (\Delta\sigma)^5$} has been found for rock salt and other materials. Therefore, an accuracy of the applied stresses of better than 1% is required. Because it is not advisable to use an ever running servovalve system for long term creep tests, one should use a mechanical system with cantilevers as shown in figure 2.4 or a hydraulic system with a load (or pressure) control which comes only into action when the load (or pressure) leaves a certain window (control on demand). Another weak point may be the stability of a computerized data collection system when a test is running over a long period of time.

It is important to use fixed platens and inflexible guides for the platens in order to avoid an inclination of the endfaces of the specimen relative to the sample axis during the tests which otherwise would be probable due to inhomogeneous deformation in the specimen. Because small inclinations are nevertheless inevitable two or three deformation transducers must be placed around the sample. Their averaged readings will further be used for determining strain. Another possibility is the use of a dilatometer system which measures the displacement between the centers of the two endfaces of the specimen. An example is shown in figure 5.1, where deformation is measured in a borehole along the axis of the specimen. Figure 5.1 shows a high precision creep testing device for uniaxial creep tests at temperatures up to 350°C which has been designed for measuring very low strain rates

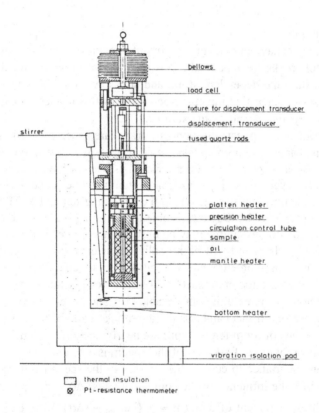

Figure 5.1: Cross section of a high-precision device for uniaxial creep tests for temperatures up to 350°C and maximum load of 95 kN (BGR, Hunsche 1988).
Strain rates down to $\dot{\varepsilon} = 10^{-12}$ s^{-1} can be mesured with this device.
Total height: 200 cm. Deformation is measured in the center along the axis of the specimen. Specimen size: 250 mm height, 100 mm diameter.

Figure 5.2: Rigs for performing uniaxial creep tests on five samples over one another with the same load (BGR). Specimen size: 250 mm height, 100 mm diameter.

down to about $\dot{\varepsilon} = 10^{-12}$ s^{-1}. Special measures had to be taken to reach this accuracy (see Geisler 1985, Hunsche 1988). Again, precise temperature control and minimization of friction between to moving parts of the rig (the two essential "enemies of precision") were most difficult. The usual limit of strain rate resolution in common rigs for creep tests is about $\dot{\varepsilon} = 10^{-10}$ s^{-1}.

Figure 5.2 shows a number of rigs which allow to carry out uniaxial creep tests simultaneously on five samples with the same load which is controlled by a precise load control (digital control on demand). Because the stress difference $\Delta\sigma$ has to be kept constant during a tests phase, the axial load has to be adjusted from time to time according to the change of the cross section due to strain by using equation (2.1), which also includes the influence of confining pressure in triaxial tests.

It is important that everybody who interprets experimental data should be aware of the experimental procedure which was used to produce these data, otherwise serious misinterpretations are possible. On the other hand, the experimental details have to be collected and reported carefully by the experimentalist.

5.2 Results of creep tests

Figure 5.3 shows the results of five simultaneous uniaxial creep tests at room temperature using specimens of the same type of rock salt (for the rig see figure 5.2). The tests have been performed with stepwise increase of the stress. One can see the subsequent tran-

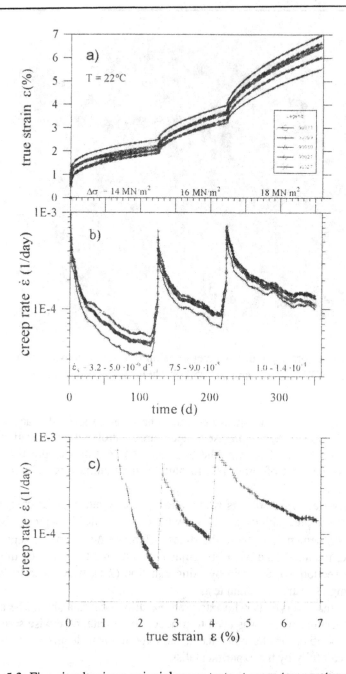

Figure 5.3: Five simultanious uniaxial creep tests at room temperature on rock salt specimens of the same type with three succeeding stresses (BGR). Rock salt (z3OS/LS) from the Asse mine (Northern Germany). Rig see figure 5.2.
a) strain vs time; b) creep rates vs time, $\dot{\varepsilon}_s$: steady state creep rates;
c) creep rate vs strain for one of the tests.

sient and steady state creep phases and the increase of steady state creep rates with increasing stress. One can also observe, that these samples behave very similarly. Often it is difficult to decide whether creep has already reached steady state. For this purpose, a plot of the creep rate as a function of strain is rather helpful as shown in figure 5.3c for one specimen. A strain of 10% is sometimes required to attain steady state creep. That is why the duration of creep tests often exceed one year.

Figure 5.4 shows the result of a uniaxial creep test carried out at four successive temperatures on a rock salt specimen. One can observe the big influence of temperature on creep rate; in practice, steady state creep rate is doubled about every 10°C. It is obvious that a temperature increase does practically not produce a transient creep phase. The same holds for a temperature decrease where one should slightly increase $\Delta\sigma$ in order to hold a constant $\Delta\sigma/G$.

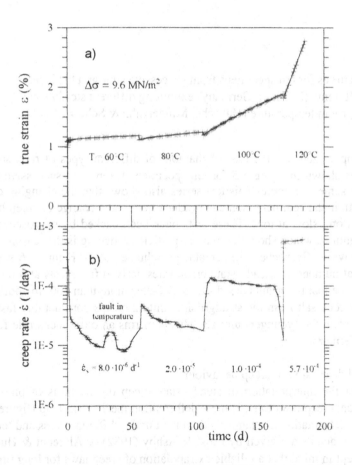

Figure 5.4: Uniaxial creep test on rock salt (z3OSU) from the Gorleben salt dome (Northern Germany) with four succeeding temperatures (BGR).
a) strain vs time; b) creep rate vs time, $\dot\varepsilon_s$: steady state creep rates.

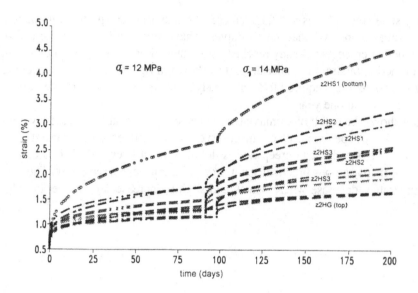

Figure 5.5: Creep curves for ten specimens from successive stratigraphic layers in the Gorleben salt dome (Northern Germany) exhibiting different steady state creep rates (BGR); room temperature (Hunsche, Mingerzahn & Schulze 1993).

It has to be emphasized that the creep behaviour of different types of rock salt can differ considerably as shown in figure 5.5 for ten specimens taken from successive stratigraphic layers of the same salt dome. This test series also shows that the changing conditions during evaporation and sedimentation can produce systematic change of creep behaviour (here: fast at bottom, slow at top). These tests have been backed by microscopic and geochemical investigations which show that mainly particle hardening is the reason for the change in creep behaviour (Hunsche, Mingerzahn & Schulze 1993). Figure 5.6 shows a compilation of a great number of stready state creep rates derived from tests at room temperature. One can notice that the results can differ by a factor more than 100. In addition, it can be observed, that rock salt from the stratigraphic unit z2 creeps considerably faster on an average than that from the (younger) unit z3. This confirms an old miner's rule for salt domes in Northern Germany.

5.3 Interpretation and discussion of creep behaviour

It is helpful for the interpretation of steady state creep on the basis of physically based micromechanical creep laws to examine a deformation mechanism map (figure 5.7). It reflects the various deformation mechanisms dominant under different stress and temperature conditions. More details are given by Frost & Ashby (1982) or Albrecht & Hunsche (1980). It has to be kept in mind that a reliable extrapolation of creep laws for long times or low strain rates, which cannot be accomplished in laboratory tests, can only be done on the basis of the knowledge of the physics of the acting deformation mechanisms on a micro-

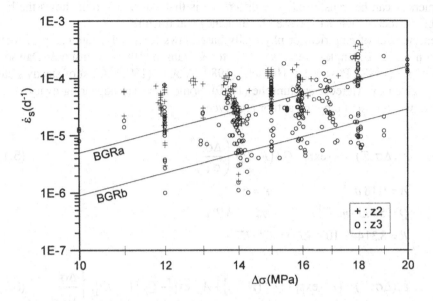

Figure 5.6: Compilation of a great number of steady state creep rates $\dot{\varepsilon}_s$ from tests at room temperature (mainly uniaxial) performed at BGR on rock salt from the Gorleben salt dome (Northern Germany). Values for $\dot{\varepsilon}_s$ differ by a factor of more than 100 .

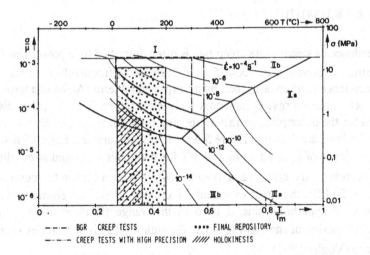

Figure 5.7: Deformation mechanism map for natural polycrystalline rock salt.
I, IIa, IIb, IIIa, IIIb: different deformation mechanisms (Hunsche 1988).

scopic basis. It has to be emphasized that the deformation mechanisms acting in rock salt are the same as in metals and ceramics (e.g. Ilschner 1973). Therefore, the knowledge of materials sciences can be transferred. The difference is that salt in addition shows the behaviour of brittle rock which is expressed by dilatancy and fracture.

A great number of empirical or physically based laws for steady state creep of rock salt have been set up. Examples are those of Carter & Hansen (1983), Munson & Dawson (1984), Wawersik (1988), Plischke & Hunsche (1989), Vogler (1992). A list of steady state creep laws is given by Albrecht & Hunsche (1980). Some more are appearing every year. As examples, two laws set up by the author are given:

BGRa: $$\dot{\varepsilon}_s(\Delta\sigma,T) = A \cdot \exp[-Q/(R \cdot T)] \cdot \left(\frac{\Delta\sigma}{\sigma_*}\right)^n \qquad (5.1)$$

$A = 0.18\ d^{-1}$ $\qquad\qquad$ $n = 5$

$Q = 54\ kJ \cdot mol^{-1}$ $\qquad\quad$ $\sigma_* = 1\ MPa$

$R = 8.31441 \cdot 10^{-3}\ kJ \cdot mol^{-1} \cdot K^{-1}$

BGRb: $$\dot{\varepsilon}_s(\Delta\sigma,T) = \left(A_1 \cdot \exp[-Q_1/(R \cdot T)] + A_2 \cdot \exp[-Q_2/(R \cdot T)]\right) \cdot \left(\frac{\Delta\sigma}{\sigma_*}\right)^n \qquad (5.2)$$

$A_1 = 2.3 \cdot 10^{-4}\ d^{-1}$ $\qquad\quad$ $A_2 = 2.1 \cdot 10^6\ d^{-1}$

$Q_1 = 42\ kJ \cdot mol^{-1}$ $\qquad\quad$ $Q_2 = 113\ kJ \cdot mol^{-1}$

$n = 5$ $\qquad\qquad\qquad$ $\sigma_* = 1\ MPa$

$R = 8.31441 \cdot 10^{-3}\ kJ \cdot mol^{-1} \cdot K^{-1}$

The stress dependence of steady state creep rate is often described by a power law $\dot{\varepsilon}_s \sim (\Delta\sigma)$; values for the stress exponent n of about 5 have been found frequently for rock salt. The temperature dependence is often described by an exponential term (Arrhenius term) were Q is the effective activation energy; Q increases with temperature. The creep law BGRa has been determined for the description of faster creeping rock salt types and BGRb for slower creeping ones. The laws are drawn in figure 5.6 for room temperature. Figure 5.8 shows the temperature dependence of $\dot{\varepsilon}_s$ as a function of 1/T for both laws at a fixed stress difference. Here, the slope represents the effective activation energy Q for a certain temperature range. Above about $\Delta\sigma = 20$ MPa the stress exponent n increases. Therefore, sometimes an exponential function or a hyperbolic sine is used in this range (Wallner 1983, Munson & Dawson 1984). A function of the form $\dot{\varepsilon}_s \sim (\Delta\sigma)^n \cdot \sinh(C \cdot \Delta\sigma)$ is sometimes used for the whole stress range (Vogler 1992).

Usually creep laws are determined for uniaxial deformation. In these cases they have to be generalized for there dimensions. In general, von Mises criterion is used for this purpose. However, it has not been ensured that this is the right criterion as it is not the right criterion for failure. To prevent the problem of generalizing Cristescu & Hunsche (1993b)

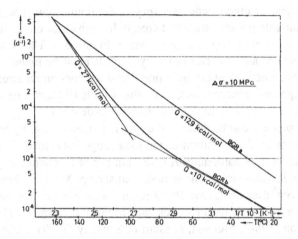

Figure 5.8: Diagram $\dot{\varepsilon}_s$ versus 1/T of the creep laws BGRa and BGRb for rock salt; see equations (5.1) and (5.2).

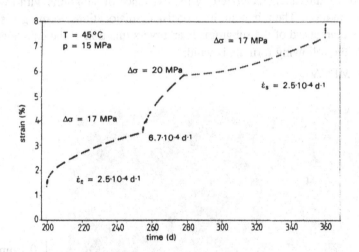

Figure 5.9: Section of a creep curve derived from a triaxial test on rock salt at elevated temperature including a phase with reduced stress (BGR). Note the invers transient creep due to recovery during this phase. $\dot{\varepsilon}_s$: steady state creep rates.

and Cristescu (1994) have shown a procedure for determining a constitutive equation for transient and steady state creep directly from the measurements.

The change of creep rate in the course of transient creep is caused by the changing relation of work hardening due to the entangling of the increasing number of dislocations and of recovery due to the annihilation of dislocations. In steady state creep these processes are in balance whereas workhardening dominates after a stress increase causing a decreasing strain rate (normal transient creep), and recovery is dominant after a stress decrease causing an increasing strain rate (inverse transient creep). In both cases creep rates are aiming at the appropriate steady state creep rate as shown in Figure 5.9. The strain activated recovery process after a stress decrease can last very long. It must be mentioned, that recovery and inverse transient creep is important in practice since continuous stress reduction is common around underground excacations due to stress redistribution (see also chapter 4.2 and 4.3). However, the invers transient creep is difficult to describe.

The transient creep process can be described by various means, e.g. by the evolution of a workhardening factor or by the evolution of the long range backstress. Some laws involve strain hardening, others time hardening. A great number of transient creep laws have been developed for rock salt, however no one is really satisfying. Many of them do not include inverse transient creep after deviatoric stress reduction, some do. The following references are given for transient creep laws for rock salt: Menzel & Schreiner (1977), Krieg (1982), Munson & Dawson (1984), Munson, Fossum & Senseny (1989), Aubertin, Gill & Ladanyi (1991), Vogler (1992), Cristescu & Hunsche (1992, 1993a,b), Cristescu (1993, 1994). A more detailed discussion will not be given here.

Steady state creep is also characterized by a (statistically) constant microscopic substructure. This substructure is characterized by the existence of subgrains with low difference in crystal orientation. Their boundaries consist of dislocations forming dislocation walls. The average diameter d of the subgrains is an invers function of the equivalent stress $\Delta\sigma$. On an average the following formula is valid:

$$d = 200 \ \mu m \cdot MPa/\Delta\sigma. \tag{5.3}$$

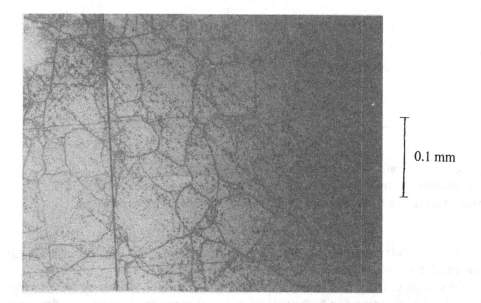

0.1 mm

Figure 5.10: Dislocations and subgrains in an etched rock salt crystal from a specimen used in a uniaxial creep test with T = 200°C and $\Delta\sigma$ = 5 MPa. d ≈ 80 μm.

This fact can be used to estimate previous tectonic stresses (Carter & Hansen 1983). The interior of the subgrains contain a relatively small number of free dislocations. Single dislocations and subgrains can be made visible by etching. Figure 5.10 shows a photo of a subgrain structure (subgrain walls and free dislocations) in rock salt. More information on this topic can be found in Eggeler & Blum (1981), Carter & Hansen (1983), Poirier (1985), Blum & Fleischmann (1988), Vogler (1992). The existence of hard regions (subgrain walls) and soft regions (subgrain interior) in a crystal has successfully been used for instance by Blum (1984) for the formulation of a comprehensive creep law for transient and steady state creep for metals and ceramics (composite model). It is obviously also efficient for rock salt (Vogler 1992).

One has to deal with the reasons for the large differences in steady state creep rates (see figures 5.5 and 5.6) as well as the differences in transient creep because they have to be taken into account in practice. For this purpose one has to distinguish between primary (microscopic, physical) parameters which influence the dominant deformation mechanisms by affecting the movement of dislocations, and secondary (macroscopic, indirect) parameters which are in some complicated way correlated with creep rates and also with the primary parameters. Primary parameters can be: impurities and chemical composition which have been analysed by Hunsche, Mingerzahn & Schulze (1993), or moisture content which can change steady state creep rates of rock salt by a factor of 50 (Hunsche & Schulze 1993). Primary parameters can be analysed by microscopic observation which is time consuming and difficult to perform. Secondary parameters can be grain size, geological situation, stratigraphy or petrography which can be determined easily; in general, they can only be analysed statistically. However, they are quite helpful because they may be used to define areas of similar creep behaviour in a salt formation. These parts of a salt dome are called "homogeneous parts". Their definition and description is important for setting up models for geomechanical model calculations (Albrecht, Hunsche & Schulze 1993).

As in failure tests, one can observe also in creep tests on rock salt that volume is slowly changing during transient and steady state creep. Examples for the occurence of volume change in creep tests is shown in figure 5.11. Volume change is caused by the opening and closing of microcracks; they can also be initiated or healed. Because only little observation of dilatancy in rock salt is available in the literature the knowledge is not very satisfying; a few examples for other geomaterials are given by Cristescu (1994, this volume). However, it is clear meanwhile that there exists a dilatancy boundary for creep of rock salt (and other geomaterials) which is very likely the same as for quasistatic tests given in figure 3.15. Figure 5.12 shows a collection of dilatancy boundaries and test results (here as function $\Delta\sigma = f(p)$) which have been determined on the basis of quasistatic tests and creep tests for rock salt. The creep tests have been performed by Spiers et al. (1989), Van Sambeck et al. (1993), and Hunsche & Schulze (1993). The results of Hunsche & Schulze mark the onset of creep acceleration due to humidity induced creep which is only present in dilatant rock salt. The relatively large differences between the curves shown in figure 5.12 are partly due to the fact that different types of rock salt have been used in the tests. However, they also indicate that further research has to be carried out in this field which is experimentally and theoretically rather complicated. One can also infer from these

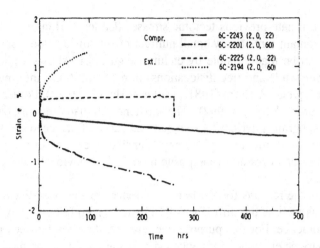

Figure 5.11: Volumetric strain in four triaxial creep tests carried out in compression or extension. T = 22 or 60°C, p = 14 MPa, Δσ = 20 MPa, e < 0: volume increase. (Wawersik, Hannum, Lauson 1980).

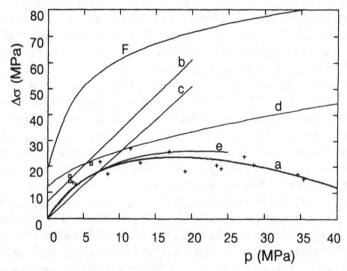

Figure 5.12: Collection of dilatancy boundaries for rock salt (boundary between the dilatant and compressible domains).

a : Cristescu & Hunsche (1993a,b),

    + : corresponding measured values (Hunsche 1992).

b : Spiers et al. (1989),

c : Van Sambeck et al. (1993),   d : Thorel & Ghoreychi (1993),

e : nonlinear fit of the data of Van Sambeck et al. (1993) by Hunsche.

F : failure surface for rock salt determined by Cristescu & Hunsche (1993b).

□ : measured values for onset of humidity induced creep (Hunsche & Schulze 1993)

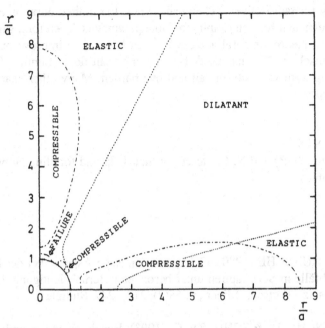

Figure 5.13: Calculated domains of failure, dilatancy, and compressibility around a horizontal tunnel in rock salt immediately after excavation on the basis of the equations plotted in figure 3.15. Elastic: Salt behaves elasticly due to deviatoric unloading.
Diameter of tunnel: a; Depth: 1000 m;
relation of horizontal stress and vertical stress: 0.3.
(Cristescu & Hunsche 1993b)

results that cracks can be healed quite easily under certain stress conditions. This knowledge is very important for the design of tight repositories.

It is an important question whether the results obtained from laboratory tests are valid also for large underground rock masses. This has to be checked by appropriate field observations and the conclusions of this comparison have to be taken into account in model calculations. Differences are present especially in jointed rock masses. For rock salt it has been shown by means of long term tests on large rock salt pillars (height: 4m, edge length: 1.5 m) in a mine that the laboratory results can be transferred to larger salt masses (Hunsche et al. 1985, Plischke & Hunsche 1989).

In situ tests and observations are also important for the validation of constitutive equations and of geomechanical models. This is often done by model calculations on boreholes, caverns, or tunnels. A number of examples are given, for instance, by Cristescu (1994, this volume). Figure 5.13 gives the result of a model calculation around a horizontal tunnel describing the dilatant and non-dilatant domains right after excavation. The time dependent evolution of the domains can be determined in additional steps.

For the prediction of the long term behaviour of structures, for instance for the safety analysis of a repository for radioactive or chemical wastes, the limited duration of laboratory tests may be not sufficient for determining reliable parameters. Therefore, suitable experiments carried out by nature - natural analogues - can be used to improve our knowledge. For instance, Hunsche (1978) has used the growth of salt dome families for determining the integral long term viscosity of salt and overburden. Many other examples are possible for similar purposes.

ACKNOWLEDGEMENTS

I thank O. Schulze (BGR) and N. Cristescu for their help and for stimulating discussions about rock mechanics.

REFERENCES

ALBRECHT, H. & U. HUNSCHE (1980): Gebirgsmechanische Aspekte bei der Endlagerung radioaktiver Abfälle in Salzdiapiren unter besonderer Berücksichtigung des Fließverhaltens von Steinsalz.- Fortschr. Miner., v. 58(2), 212 - 247. Stuttgart.

ALBRECHT, H., HUNSCHE, U. & SCHULZE, O. (1993): Results from the application of the laboratory test program for mapping homogeneous parts in the Gorleben salt dome.- Geotechnik-Sonderheft, 10. Nat. Felsmech. Symp., Aachen 1992, mit Beiträgen vom 7. Int. Felsmech. Kongress, Aachen 1991.- p. 152 - 155.

AUBERTIN, M., D. E. GILL & B. LADANYI (1991): A unified viscoplastic model for the inelastic flow of alkali halides.- Mech. of Materials, v. 11, 63 - 82.

BLUM, W. (1984): On the evolution of the dislocation structure during work hardening and creep.- Scripta metall., v. 18, 1383.

BLUM, W. & C. FLEISCHMANN (1988): On the deformation mechanism map of rock salt.- In: The Mechanical Behavior of Salt II, Proc. of the Second Conf., Hannover (FRG) 1984; Editors: H. R. Hardy Jr. & M. Langer; p. 7 - 22. Trans Tech Publications, Clausthal.

BLUM, W. & F. PSCHENITZKA (1976): Durchführung von Zug-, Druck- und Spannungsrelaxationsverfahren bei erhöhter Temperatur.- Z. Metallkunde, Bd. 67, Heft 1, 62 - 65.

BRACE, W. F. (1964): Brittle fracture of rocks.- In State of Stress in the Earth's Crust, Proc. of the Int. Conf., Santa Monica (USA) 1963; Editor: W. R. Judd, p. 111 - 178. Am. Elsevier, New York.

CARTER, N. L. & F. D. HANSEN (1983): Creep of rocksalt.- Tectonophys., v. 92, 275 - 333.

CARTER, N. L., S. T. HORSEMAN, J. E. RUSSELL & J. HANDIN (1993): Rheology of rocksalt.- J. Structural Geology, v. 15, 1257 - 1271.

CARMICHAEL, R. S. (1982, 1984): CRC, Handbook of physical properties of rocks.- Vol. 1 & 2 (1982), Vol. 3 (1984). CRC Press, Boca Raton, Fla.

CLARK Jr., S. P. (1966): Handbook of Physical Constants.- Geol. Soc. of Am., Memoir 97. New York.

CRISTESCU, N. (1989): Rock Rheology.- 336 p., Kluwer Acad. Publ., Dordrecht.

CRISTESCU, N. (1993): Constitutive equation for rock salt and mining application.- In: H. Kakihana, H. R. Hardy, Jr., T. Hoshi, & K. Toyokura (Editors), Proc. Seventh Symp. on Salt, Kyoto (Japan) 1992, v. I, p. 105 - 115, Elsevier, Amsterdam.

CRISTESCU, N. (1994): Viscoplasticity of geomaterials.- In: Time-dependent behaviour of geomaterials. CISM-course, Udine. Springer, Wien. (This volume)

CRISTESCU, N. & U. HUNSCHE (1992): Determination of a nonassociated constitutive equation for rock salt from experiments.- In: Finite Inelastic Deformations - Theory and Applications, Proc. IUTAM-Symposium Hannover, Germany 1991; Editors: D. Besdo & E. Stein; p. 511-523. Springer, Berlin.

CRISTESCU, N. & U. HUNSCHE (1993a): A constitutive equation for salt.- In: Proc. 7. Int. Congr. on Rock Mech., Workshop on Rock Salt Mech., Aachen, Sept. 1991, v. 3, p. 1821 - 1830. Balkema, Rotterdam.

CRISTESCU, N. & HUNSCHE, U. (1993b): A comprehensive constitutive equation for rock salt: determination and application.- In: The Mechanical Behavior of Salt III, Proc. of the Third Conf., Palaiseau (France) 1993.- Clausthal (Trans Tech Publications). (in press.)

DREYER, W. (1972): The Science of Rock Mechanics. Part 1: The Strength Properties of Rocks.- 501 p., Trans Tech Publications, Clausthal.

EGGELER, G. & W. BLUM (1981): Coarsening of the dislocation structure after stress reduction during creep of NaCl single crystals.- Philos. Mag., A44, 1065 - 1084.

FROST, H. J. & M. F. ASHBY (1982): Deformation-Mechanism Maps. The Plasticy and Creep of Metals and Ceramics.- 164 p., Pergamon Press, Oxford.

GEISLER, G. (1985): Entwicklung einer Präzisionsapparatur zur Messung des Kriechverhaltens von Steinsalzproben.- Dissertation, 123 p., Universität Hannover.

GEVANTMAN, L. H. (1981): Physical Properties Data for Rock Salt.- NBS Monograph 167, 282 p., U.S. Department of Commerce/National Bureau of Standards.

HARDY Jr., H. R. & M. LANGER (1984): The Mechanical Behavior of Salt.- Series on Rock and Soil Mechanics, v. 4. Proc. of the First Conf., University Park (USA) 1981. 901 p., Trans Tech Publications, Clausthal.

HARDY Jr., H. R. & M. LANGER (1988): The Mechanical Behavior of Salt II.- Series on Rock and Soil Mechanics, v. 14. Proc. of the Second Conf., Hannover (FRG) 1984. 781 p., Trans Tech Publications, Clausthal.

HAUPT, M. (1988): Entwicklung eines Stoffgesetzes für Steinsalz auf der Basis von Kriech- und Relaxationsversuchen.- Dissertation. Veröffentlichungen des Inst. für Bodenmech. u. Felsmech. der Univ. Karlsruhe (Germany). Editors: G. Gudehus & O. Natau; Heft 110, 138 p., Karlsruhe.

HERTZBERG, R.W. (1983): Deformation and fracture mechanics of engineering materials.- 697 p., John Wiley & Sons, New York.

HEEMANN, U. (1989): Transientes Kriechen und Kriechbruch im Steinsalz.- Dissertation, Univ. Hannover. Forschungs- und Seminarberichte aus dem Bereich der Mechanik der Universität Hannover. 102 p., Bericht-Nr. F 89/3.

HUNSCHE, U. (1978): Modellrechnungen zur Entstehung von Salzstockfamilien.- Geol. Jb., E 12, p. 53 - 107. Hannover.

HUNSCHE, U. (1988): Measurements of creep in rock salt at small strain rates.- In: The Mechanical Behavior of Salt II, Proc. of the Second Conf., Hannover (FRG) 1984; Editors: H. R. Hardy Jr. & M. Langer; p. 187 - 196. Trans Tech Publications, Clausthal.

HUNSCHE, U. (1992): True triaxial failure tests on cubic rock salt samples - experimental methods and results.- In: Finite Inelastic Deformations - Theory and Applications, Proc. IUTAM-Symposium Hannover, Germany 1991; Editors: D. Besdo & E. Stein; p. 525-536. Springer, Berlin.

HUNSCHE, U. (1993): Failure behaviour of rock salt around underground cavities.- In: H. Kakihana, H. R. Hardy Jr., T. Hoshi, & K. Toyokura (Editors), Proc. Seventh Symp. on Salt, Kyoto (Japan). April 1992, v. I, p. 59 - 65, Elsevier, Amsterdam.

HUNSCHE, U. & H. ALBRECHT (1990): Results of true triaxial strength tests on rock salt.- Engineering Fracture Mechanics, v. 35, No. 4/5, 867 - 877.

HUNSCHE, U., I. PLISCHKE, H.-K. NIPP & H. ALBRECHT (1985): An in situ creep experiment using a large rock salt pillar.- In: Proc. Sixth Int. Symp. on Salt, Toronto 1983; Editors: B. C. Schreiber & H. L. Harner; v. I, p. 437 - 454. Alexandria (USA) (The Salt Institute).

HUNSCHE, U. & O. SCHULZE (1993): Effect of humidity and confining pressure on creep of rock salt.- In: The Mechanical Behavior of Salt III, Proc. of the Third Conf., Palaiseau (France) 1993. Trans Tech Publications, Clausthal (in press).

HUNSCHE, U., G. MINGERZAHN & O. SCHULZE (1993): The influence of textural parameters and mineralogical compostion on the creep behavior of rock salt.- In: The Mechanical Behavior of Salt III, Proc. of the Third Conf., Palaiseau (France) 1993. Trans Tech Publications, Clausthal (in press).

ILSCHNER, B. (1973): Hochtemperatur-Plastizität.- 314 p., Springer, Berlin.

JAEGER, J. C. & N. G. W. COOK (1979): Fundamentals of rock mechanics.- 593 p., Chapman and Hall, London.

JUMIKIS, A. R. (1983): Rock Mechanics.- 613 p., Trans Tech Publications, Clausthal.

KLEPACZKO, J. R., G. GARY & P. BARBERIS (1991): Behaviour of rock salt in uniaxial compression at medium and high strain rates.- Arch. Mech., v. 43, 499 - 517.

KRIEG, R. D. (1982): A unified creep-plasticity model for halite.- Special Technical Publ. 765, 139 - 147, ASTM, Philadelphia.

LAMA, R & V. VUTUKURI (1974/78): Handbook on Mechanical Properties of Rocks - Testing Techniques and Results.- Vol. I, II, III, IV; 1746 p., Trans Tech Publications, Clausthal.

LANDOLT-BOERNSTEIN (1982): Physikalische Eigenschaften der Gesteine.- Gruppe 5, Band 1, Teilband b. Editor: G. Angenheister. Springer, Heidelberg.

LANGER, M. (1986): Rheology of rock-salt and its application for radioactive waste disposal purposes.- Proc. Int. Symp. on Engineering in Complex Rock Formations, Peking 1986.- p. 1 - 19.

LLIBOUTRY, L. (1987): Very Slow Flows of Solids.- 510 p., Martinus Nijhoff Publ., Dordrecht.

MECKING, H. & Y. ESTRIN (1987): Microstructure-related constitutive modelling of plastic deformation.- In: Proc. 8th Risø International Symposium of Metallurgy and Materials Science "Constitutive Relations and Their Physical Basis", Risø National Laboratory, Roskilde, Denmark, 1987; Editors: S. I. Andersen et al.; p. 123 - 145.

MENZEL, W. & W. SCHREINER (1977): Zum geomechanischen Verhalten von Steinsalz verschiedener Lagerstätten der DDR. Teil II: Das Verformungsverhalten.- Neue Bergbautechnik, v. 7(8), 565 - 571.

MUNSON, D. E. & P. R. DAWSON (1984): Salt constitutive modeling using mechnism maps.- In: The Mechanical Behavior of Salt, Proc. of the First Conf., University Park (USA) 1981; Editors: Hardy, H. R. Jr. & M. Langer; p. 717 - 737. Trans Tech Publications, Clausthal.

MUNSON, D., A. FOSSUM & P. SENSENY (1989): Advances in resolution of discrepancies between predicted and measured in situ WIPP room closures.- SAND88-2948, Sandia National Laboratories, Albuquerque, USA (NM).

PATERSON, M. S. (1978): Experimental Rock Deformation - The Brittle Field.- 254 p., Springer-Verlag, Berlin.

PLISCHKE, I. & U. HUNSCHE (1989): In situ-Kriechversuche unter kontrollierten Spannungsbedingungen an großen Steinsalzpfeilern.- In: Rock at Great Depth, Proc. ISRM-SPE Int. Symp., Aug.- Sept. 1989, Pau (France); Editors: V. Maury & D. Fourmaintraux; vol. 1, p. 101 - 108. A.A.Balkema, Rotterdam.

POIRIER, J.-P. (1985): Creep of crystals. High-temperature deformation processes in metals, ceramics and minerals.- 260 p. Cambridge Univ. Press, Cambridge.

SPIERS, C. J, P. M. SCHUTJENS, R. H. BRZESOWSKJ, C. J. PEACH, J. L. LIEZEN-BERG & H. J. ZWART (1990): Experimental determination of constitutive parameters governing creep of rocksalt by pressure solution.- In: Deformation Mechanisms, Rheology and Tectonics; Editors: Knipe, R. J. & E. H. Rutter; Geological Special Publication No. 54, p. 215 - 227.

VAN SAMBECK, L., A. FOSSUM, G. CALLAHAN & J. RATIGAN (1993): Salt mechanics: Empirical and theoretical developments.- In: H. Kakihana, H. R. Hardy, Jr., T. Hoshi, K. Toyokura (Editors), Proc. Seventh Int. Symp on Salt, Kyoto (Japan). April 1992, v. I, p. 127 - 134. Elsevier, Amsterdam.

VOGLER, S. (1992): Kinetik der plastischen Verformung von natürlichem Steinsalz und ihre quantitative Beschreibung mit dem Verbundmodell.- Dissertation, 148 p., Universität Erlangen-Nürnberg.

WALLNER, M. (1983): Stability calculations concerning a room and pillar design in rock salt.- In: Proc. 5th Int. Congr. on Rock Mechanics, Melbourne 1983, v. II, p. D9 - D15, A. A. Balkema, Rotterdam.

WAWERSIK, W. (1988): Alternatives to a power-law creep model for rock salt at temperatures below 160°C.- In: The Mechanical Behavior of Salt II, Proc. of the Second Conf., Hannover (FRG) 1984; Editors: H. R. Hardy Jr. & M. Langer; p. 103 - 128. Trans Tech Publications, Clausthal.

WAWERSIK, W. & D. HANNUM (1980): Mechanical behavior of New Mexico rock salt in triaxial compression up to 200°C.- J. Geophys. Res., B85, 891 - 900.

WAWERSIK, W., D. HANNUM & H. LAUSON (1980): Compression and extension data for dome salt from West Hackbury, Loisiana.- Sandia-Report SAND79-0668, 35 p., Sandia National Laboratories, Albuquerque NM, USA.

EXPERIMENTAL EVIDENCE OF NON-LINEAR
AND CREEP BEHAVIOUR OF PYROCLASTIC ROCKS

A. Evangelista and S. Aversa
University of Naples Federico II, Naples, Italy

Abstract: An analysis of the mechanical behaviour of pyroclastic rocks is discussed in this paper, with particular evidence to non-linear and creep behaviour. This topic is treated presenting experimental results obtained on Neapolitan tuffs at the University of Naples. Standard tests, which outline the non-linear behaviour of these rocks, have been performed for many years and their results have been presented in many papers. Although creep tests on the Neapolitan tuffs have also been performed for many years, their results have never been published before this paper. For this reason a substantially different approach is used in discussing the non-linearity and creep behaviour of these materials. In the paper a comparison with comparable behaviour of other rocks and soils is presented.

1. Introduction

The mechanical behaviour of soft rocks is intermediate between that of soils and that of hard rocks. A simple criterion used to distinguish soft rocks from soils and hard rocks refers to the uniaxial strength of the material. The upper limit of the strength is often assumed equal to 25 MPa [1]. Different values of the lower limit have been proposed in the scientific literature, among which the value of 1.2 MPa, proposed by the Geological Society of London [2], seems to be the most convenient.

The uniaxial compressive strength of many pyroclastic rocks falls into this range of strength. In Fig. 1 the uniaxial strength of many soft rocks and cohesive soils is plotted vs. the porosity of the material [3]. It is also evident that the porosity is only one of the factors influencing the strength.

In recent years many researches have been addressed to the study of the mechanical behaviour of soft rocks, both natural or artificially reconstituted in the laboratory. In particular a large number of investigations have been carried out on pyroclastic rocks in Japan.

Pyroclastic rocks are widely diffused in Naples and in its surroundings (Neapolitan Yellow Tuff, Campanian Grey Tuff. etc.). For this reason many studies on these materials have been carried out by researchers of the Istituto di Tecnica delle Fondazioni e Costruzioni in Terra of the University of Naples for more than twenty years.

The present paper is mainly based on experimental results obtained during these investigations. When possible and appropriate a comparison with comparable behaviour of other rocks and soils is presented.

The paper is divided into two parts: one deals with the non-linear behaviour and the other with the creep behaviour of tuffs. The two parts are treated with a substantially different approach. The non-linear behaviour of the Neapolitan tuff and other volcanic rocks, obtained by means of traditional techniques, has been presented in many papers and therefore only its pattern of behaviour is discussed. Instead, creep test results on the Neapolitan tuffs have never been published and so in the second part of the paper the focus is on the test results and their interpretation.

In the following the mean total stress p, the mean effective stress p′ and the deviatoric stress q for axial - symmetric state of stress are expressed, according to the Cambridge convention, as:

$$p = \frac{\sigma_1 + 2\sigma_3}{3}$$

$$p' = \frac{\sigma_1' + 2\sigma_3'}{3} \tag{1}$$

$$q = \sigma_1 - \sigma_3$$

The volumetric and the distorsional components of strains are consequently expressed as:

$$\varepsilon_v = \varepsilon_1 + 2\varepsilon_3$$

$$\varepsilon_s = \frac{2}{3}(\varepsilon_1 - \varepsilon_3) \tag{2}$$

Compressive stresses and contractive strains are assumed as positive.

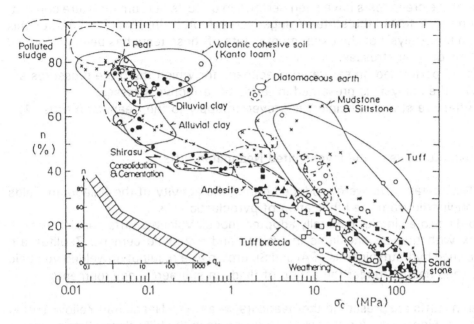

Fig. 1 - Relationship between porosity and uniaxial compressive strength (after Ogawa [3])

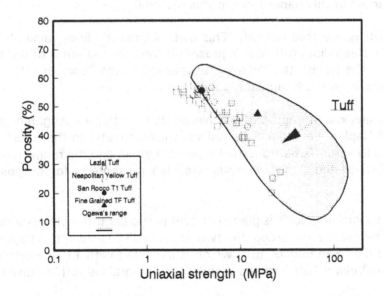

Fig. 2 - Italian tuffs: relationship between porosity and uniaxial compressive strength

Most of the creep tests have been performed on tuffs at room moisture content. During these tests the state of negative pore pressure is unknown. For this reason the analysis of the experimental data of these tests has been performed in terms of total stresses.

For tests performed on saturated specimens for which the pore pressures are known, the analysis is presented in terms of effective stresses.

The effective stress principle has proven to apply to the Neapolitan tuffs [4].

2. Characteristics of the investigated Neapolitan tuffs

The Neapolitan tuffs were produced during the activity of the Phlegrean Fields and originated from the lithification of pyroclastic soils.

According to a simple, generally adopted model, volcanic tuffs consist of ashy matrix with pumice and lithic inclusions and a zeolitic cement. Zeolites are neoformed hydrate minerals of Al and Si, grown as a result of unwelded volcanic glass modification under the action of fluids in sub-aereal environments.

Different tuffs are present in the Neapolitan area. The Neapolitan Yellow Tuff is the most widespread formation which outcrops in the hilly slopes that surround Naples, while in the flatter areas it is located a few meters below ground level. Most of the formation is above the water table.

This tuff has been widely investigated at the University of Naples. Many of the results presented in this paper refer to this material.

A different tuff was tested recently. This material, called "fine-grained tuff", is older than the Neapolitan tuff and is present in very limited areas of the city in pockets under the Neapolitan Yellow Tuff packed in the "ancient tuffs". This material differs from the Neapolitan Yellow Tuff.

In Fig. 2 the uniaxial strengths of the Neapolitan Tuffs, investigated at the University of Naples, have been plotted vs. their porosity. In the figure some points related to some Tuffs from the Region of Latium and the range identified by Ogawa [3] for the Japanese tuffs are also presented for purposes of comparison.

In the figure special emphasis is placed on two particular tuffs which have been tested for non-linearity and creep, the two aspects discussed in this paper. The two tuffs are the "San Rocco" tuff, which is indicated with T1 and belongs to the Neapolitan Yellow Tuff formation, and the "fine-grained tuff" referred to as TF.

Table 1 summarises some of the physical and mechanical properties of these two tuffs.

The main difference between the two tuffs concerns the pores. They are uniform and small in the TF tuff, and irregular in shape and larger in the T1 tuff. This feature justifies the different values of the permeability coefficient.

Pumice inclusions prevail in the T1 tuff , they are welded in the outer zone and unwelded inside. The arrangement of pumices is nearly always chaotic. In the TF tuff only a few, very small lithic inclusions are present.

The T1 Tuff, like other kinds of the Neapolitan Yellow Tuff, is characterised by a high variability in structure that is matched by a wide variability in the mechanical behaviour.

Table. 1 - Main physical and mechanical properties of T1 and TF tuffs

Symbol	Tuff	Specific gravity γ_s (kN/m³)	Porosity n (%)	Dry unit weight γ_d (kN/m³)	Uniaxial Compressive strength σ_c (MPa)	Permeabilit y coefficient k (cm/s)
T1	San Rocco	24,4	55,5	10,85	3,88	1,5E-6 6,4E-6
TF	Fine grained	24,30	47,30	12,80	15,29	1,6E-7

The compressive strength refers to tuff dried at atmospheric humidity, and tested at the standard strain rate

Relevant scatters have been noticed also in the TF tuff which had been considered as being more homogenous according to the initial tests.

This remark is supported by the variability coefficient of the uniaxial compressive strength that is high for all tuffs and is close to 20%.

Sometimes the data scatter makes the numerical evaluation of the influence of a single parameter difficult and only permits the identification of a trend. This aspect takes on particular importance in the study of creep behaviour.

Feda [5] gives some indication of the variability of the creep deformation of different materials generally considered to be homogeneous: metallic specimens of the same melt differ by up to 20%; similar values are reached by aluminium and concrete. Only polymers have a considerable reduction in scatter equal to ±5%.

On this basis it is interesting to stress that the expected scatter of the creep phenomenon of tuff will be greater than the intrinsic variability of the rock.

3. Deformability

As already mentioned, the mechanical behaviour of tuff has been widely investigated at the Istituto di Tecnica delle Fondazioni e Costruzioni in Terra of the University of Naples. During a first period investigations concerned the

Neapolitan Yellow Tuff [6, 7, 8, 9, 10, 11, 12, 13, 14, 15, 16, 17, 4].
Evangelista and Pellegrino [18] presented a comprehensive report of the physical
and mechanical properties of the Neapolitan Yellow Tuff and other Italian soft
rocks. More recently interest has been focused on the Fine-Grained Tuff
described above [19, 20, 21, 22].
The purpose of this paragraph is to describe the non-linear behaviour of volcanic
tuffs, with more attention being paid to the pattern than to a quantitative
analysis of the physical and mechanical properties. For quantitative details it is
possible to refer to the above-mentioned publications, and especially to
Evangelista and Pellegrino [18] for different types of the Neapolitan Yellow Tuff,
and to Aversa, Evangelista and Ramondini [20] for the Fine-Grained Tuff.

3. 1. Isotropic and oedometric compression

The non-linear behaviour of tuff was observed for the first time by Pellegrino [6],
who stressed that the compressibility of the Neapolitan Yellow Tuff in isotropic
compression tests is highly influenced by the mean stress values.
At low stress levels the material shows small and approximately reversible
volumetric strains (see Fig. 3). The material has a "rock-like" behaviour. An

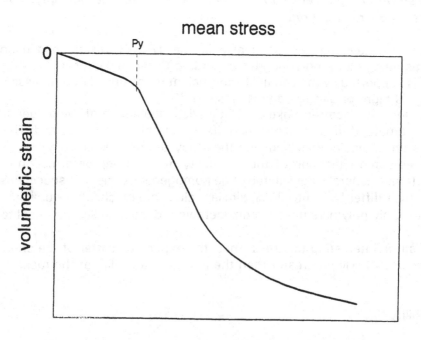

Fig. 3 - Schematic pattern of the relationship between volumetric strain and
isotropic stress for Neapolitan Yellow Tuff

elastic model can be fruitfully used in this range of stresses. At higher values of isotropic stress the material shows higher and essentially non-reversible volumetric strains. The material behaves like a soil. Simple elastic models are no longer suitable for this stress range: the dissipative nature of the strain must be taken into account.

Obviously, the transition between "rock-like" and "soil-like" behaviour is not abrupt. It is connected with the growth of a micro crack caused by a gradual breakage of bonds between individual particles and of the particles themselves. According to Leroueil and Vaughan [23] this process is called "destructuration". A more accurate description of such behaviour is provided by Aversa, Evangelista and Ramondini [20] for Fine-Grained Tuff (TF) under both isotropic and oedometric conditions. The results of three isotropic compression tests on this tuff saturated with water are shown in a conventional semi-logarithmic void ratio e vs. isotropic effective stress p' diagram (Fig. 4a) and in a bulk modulus K vs. isotropic effective stress p' diagram (Fig.4b).

The material behaviour is "rock-like" in the first part of the stress-strain curve: the bulk modulus is in fact approximately constant. At higher stresses a dramatic decrease in stiffness, connected to the destructuration phase, is observed in Fig. 4b. A gradual increase in stiffness subsequently occurs, as for "non-structured" soils.

The same kind of results have also been obtained in oedometric compression tests performed on the same tuff. The results of one of these tests are shown in Fig.5a. Observations similar to those made for the isotropic tests apply also to the oedometric test results. The first part of the diagram of the oedometric modulus from point A' to B' in Fig. 5b, in which stiffness increases, is probably due to the non-perfect initial contact between the tuff specimen and the containing ring.

Unloading and reloading cycles, performed during isotropic and oedometric tests, show the non-reversible nature of the strains and the stiff behaviour of the material during these cycles. In Fig. 5c elastic deformations and total deformations are shown.

The representation of the bulk and oedometric moduli (see Fig. 4b and 5b) gives a clear picture of the destructuration phenomenon. Most of the bond and particle breaking develops in the stress range that goes from point B' to point C'.

The destructuration is not a continuous process but seems to be a stepwise adaptation of the material structure to the load, as shown by undulation and rippling in the curve reported in Fig. 5b. Such a phenomenon is similar to the structural perturbation observed in creep curves by Conrad [24] and Lubahn [25].

The destructuration process pertains also to other soft rocks and structured soils. Examples of this kind are the collapsible undisturbed loess [5], the Leda

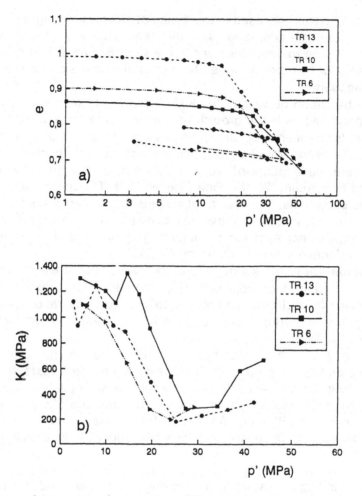

Fig. 4 - Isotropic compression tests on TF tuff: a) void ratio vs. stress; b) bulk modulus vs. stress

Clay [26] and many other soils as shown by Leroueil and Vaughan [23].
It is interesting to notice that, in both isotropic and one-dimensional compression test diagrams, the Fine-Grained Tuff, other Neapolitan tuffs and other soft rocks [27] do not show the typical shape with a point of inflection as observed for many structured soils [23].

The values of isotropic compression yield stress are generally related to the strength of the tuff. For the San Rocco tuff (T1)and the Fine-Grained tuff (TF) this value is equal to $1.5 \div 2.0$ times the uniaxial compressive strength. A different ratio, approximately equal to 1.0, has been obtained for another Yellow Tuff, the " Quarto Tuff".

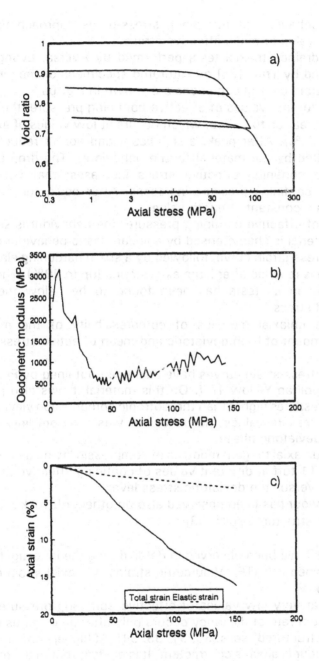

Fig. 5 - Oedometric compression test on TF tuff: a) void ratio vs. vertical stress; b) oedometric modulus vs.axial stress; c) total and elastic strains vs. axial stress

3. 2. Triaxial compression

The mechanical behaviour of tuff along stress-paths approaching failure is influenced by the value of mean effective stress.

Some results of drained triaxial tests performed by Aversa, Evangelista and Ramondini [20] and by Travi [28] on saturated specimens of the Fine-Grained Tuff (TF) under strain control conditions are shown in Fig. 6a.

It is evident that, for low values of effective confining pressure, the behaviour is approximately linear up to failure, which occurs at low values of axial strains (approximately 0.5 %). After peak, a significant and abrupt reduction of the strength is exhibited by the material (brittle behaviour). The drop in strength decreases as the confining effective stress increases; that is, brittleness decreases as the confining pressure increases. After peak, the stress-strain curve approaches a constant strength value.

At higher values of effective confining pressure, the behaviour is significantly different. The material is characterised by a linear-elastic behaviour only in the first part of the stress-strain curve, followed by a strain hardening elasto-plastic behaviour. Failure is reached after large axial strains (up to 20%). The effective confining pressure in all tests has been found to be below the isotropic compression yield stress.

Before yield, the uniaxial modulus of compressibility of the materials is practically independent of both deviatoric and mean effective stresses.

Similar trends in stress-strain curves have also been obtained by Pellegrino [7, 8, 9] on the Neapolitan Yellow Tuff. On this material it was also possible to apply confining pressures higher than the isotropic compression yield stress. For these confining pressure values the behaviour was far from linear from the beginning of the deviatoric phase.

The values of the uniaxial tangent modulus of compressibility measured on some specimens of the T1 tuff at different values of confining effective pressure are reported in Fig. 7 versus the deviatoric stress level.

This kind of behaviour has been observed also on other soft rocks [29, 30, 31, 32], and on some structured soils [33].

Volumetric strains have been observed in detail during the investigation on the saturated fine-grained tuff (TF). Volumetric strains vs. axial strain curves are reported in Fig.6b

Tuff dilates only at very low values of effective confining pressure (below 1 MPa). The maximum rate of dilatancy occurs only after peak, as usual for soft rocks and many "structured" soils [30, 31, 33, 34]. At higher confining pressure values, the behaviour is always contractant. It is evident that in an intermediate range of confining pressure the experimental data show a contractant behaviour associated with a well-defined brittle behaviour.

In other soft rocks, characterised by low porosity as the calcarenite tested by

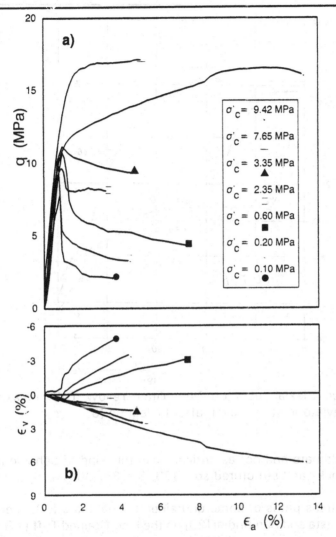

Fig. 6 - Drained triaxial tests on TF tuff: a) deviatoric stress vs. axial strain; b) volumetric strain vs. axial strain

Elliot and Brown [30], the transition between dilatant and contractant behaviour coincides with the change between strain softening and strain hardening behaviour.

On the Neapolitan Yellow Tuff a dilatant behaviour was observed only in some of the specimens of the San Rocco tuff (T1) tested under uniaxial compression. In the triaxial tests the Neapolitan Yellow Tuff, in the range of explored confining stresses, has always shown a contractant behaviour.

Coming back to the TF tuff (see Fig. 6), for high axial strains (up to 15-20%) strength assumes a practically constant value but the volumetric strain is still not stabilised. For this reason Aversa, Evangelista and Ramondini [20] defined

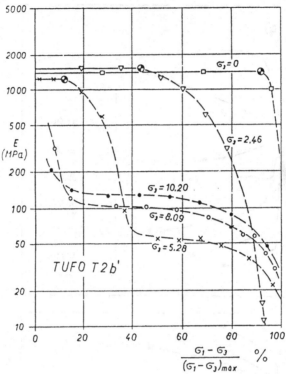

Fig. 7 - Drained triaxial tests on Neapolitan Yellow Tuff: uniaxial tangent modulus vs. deviatoric stress level (after Pellegrino [8])

this state as ultimate and not as critical. Also this kind of behaviour is typical of many soft rocks and structured soils [33, 34, 35, 36].

The effective stress paths of some undrained triaxial tests (CIU) performed by Aversa, Evangelista and Ramondini [20] on the Fine-Grained Tuff (TF) are shown in Fig.8. The specimen consolidated at an intermediate value of effective confining pressure shows, after peak, a dramatic reduction in strength due to a significant increase in pore pressure. This behaviour can be explained considering that this test has been performed in the stress range for which the material shows, in drained tests, a brittle and contractant behaviour. The specimen consolidated at low values of effective confining pressure shows, after peak, a small decrease in strength, connected to a decrease in pore pressure, due to the tendency to dilate, typical of this stress-range. The specimen consolidated at higher values of effective confining pressure shows a stress-path typical of slightly overconsolidated soils, that agrees with drained tests performed in this stress range.

Some undrained tests performed on a Japanese volcanic tuff by Akai et al.[37] and on a Canadian sensitive clay by Lefebvre [33] show a similar pattern.

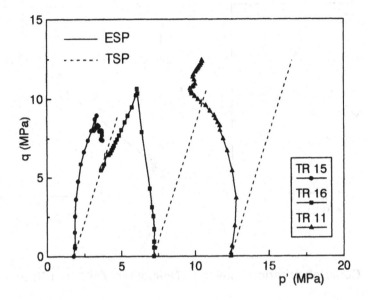

Fig. 8 - Undrained triaxial tests on TF tuff: effective stress-paths

4. Failure and yield

The strength of the Neapolitan Yellow Tuff has been usually evaluated by means of the Mohr-Coulomb failure criterion, by interpolating the maximum deviatoric stresses measured in triaxial tests (see Fig.9). Other possible criteria, like the Griffith parabola, have also been tested by Pellegrino [8, 9], who proved that this criterion does not apply correctly to the tuff.

On the basis of recent findings [38], Aversa, Evangelista and Ramondini [20] distinguished peak strength from ultimate strength of the TF tuff and used a failure envelope composed of two parts: the first interpolates the peak stresses, while the second interpolates the ultimate stresses. According to Adachi et al.[38], peak and ultimate strengths are represented by two power functions (see Fig. 10a):

$$\frac{q}{p_1} = \alpha \left(\frac{p'}{p_1}\right)^\beta \tag{3}$$

in which p_1 is an arbitrary reference pressure and α and β are coefficients that assume different values for peak and ultimate strength that depend on the

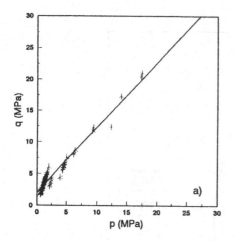
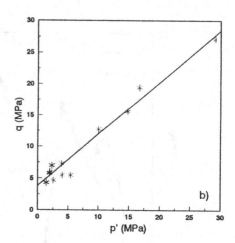

Fig. 9 - Mohr - Coulomb criterion applied to Neapolitan Yellow Tuff: a) T1 tuff; b) "Quarto" tuff

choice of reference pressure p_1.

According to the Mohr Coulomb criterion two linear functions, one for the peak and the other for the ultimate conditions, have also been used (see Fig. 10b). The Adachi et al. [38] criterion follows the experimental data much closer, showing a better agreement especially at low values of the mean effective stress.

Both criteria present some problems. The first criterion does not consider the possibility of defining a uniaxial tensile strength, passing the peak strength envelope through the origin of the axis. The bi-linear criterion predicts a meaningless cohesion intercept for ultimate strength.

The Adachi et al. [38] criterion may be modified in order to take into account the possibility for the material to sustain tensile stresses [28]. The new failure criterion is expressed by the following equation:

$$\frac{q}{p_1} = \alpha \left(\frac{p' + p'_t}{p_1}\right)^\beta \qquad (4)$$

in which p'_t is a pressure that assumes a null value for ultimate strength and a non-zero value for peak strength. The value of p'_t can be obtained with a trial and error procedure. The interpolation of peak and ultimate strength of the TF tuff with this criterion is shown in Fig. 11. Results of two tensile tests have also been used in order to obtain the value of coefficient p'_t.

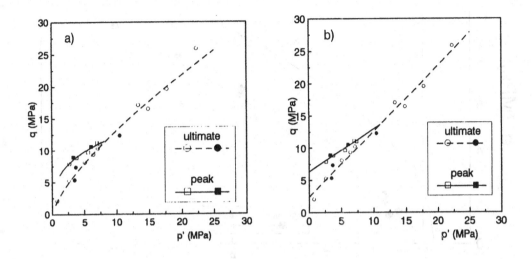

Fig. 10 - TF tuff failure envelope: a) Adachi et al. criterion; b) bilinear criterion (full symbols apply to CIU tests)

The peak strength of "structured" soils has been studied by some researchers [39, 40, 41], who experimented on artificially cemented soils. On this aspect two different ideas have been proposed. Clough et al. [39] and Maccarini [40] believe that inter-particle bonds induce only a cohesion increase in peak strength. On the other hand, comparing the peak strength of cemented and uncemented soils Lade and Overton [41] noticed that the presence of bonding induces an increase in both cohesion and friction angle. The increase in the friction angle is justified by an increase in dilatancy related to the effective size of soil particles, that in cemented soils are composed of smaller particles cemented together. Both ideas are probably correct, the first being valid for high porosity materials and the second for low porosity materials. With regard to the TF tuff the peak strength is mainly due to an increase in cohesion, and the material exhibits a dilatant behaviour only in the very low stress range and exclusively after peak [22].

The ultimate strength envelope of the TF tuff has a curve shape (see Fig.10a and Fig.11). This feature has also been noticed by Lefebvre [33] with reference to sensitive clays. According to Aversa et al. [22] this feature was due to the presence of dilatancy (at low values of the mean effective stress) and contractancy (at high values of the mean effective stress) in correspondence to the ultimate state condition.

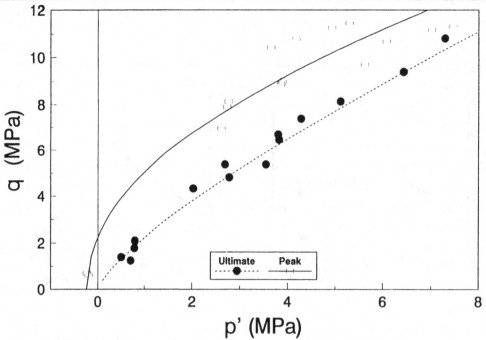

Fig. 11 - TF tuff failure envelope: criterion described by Eq.(4)

As noticed before, two different types of mechanical behaviour can be identified: a "rock-like" and a "soil-like" behaviour. The passage from one of these behaviours to the other, corresponding to the beginning of the destructuration, has been identified by Pellegrino [8, 9] with a curved line (see Fig. 12).

The similarity of this curve with the cap yield curves of some of the constitutive models developed for soils and the presence of a peak strength for stress states contained in the "rock-like" region induced many researchers, interested in modelling the behaviour of soft rocks, to use yield criteria derived from Critical State Soil Mechanics [30, 31, 35, 36, 42, 43, 44, 45, 46].

Aversa, Evangelista and Ramondini [20], with regards to TF tuff, tested two different yield criteria, derived respectively from Cam-Clay and modified Cam-Clay models, modified in order to take into account the uniaxial tensile strength of the material (see Fig.13). The arrows reported in the figure represent the direction of plastic strain increment measured in correspondence to yield stress obtained from CID triaxial tests. The direction of plastic strain vector in isotropic compression tests is parallel to the x-axis. The determination of plastic strain vectors from undrained test is not so reliable, requiring an accurate determination of elastic volumetric strains. For this reason the directions of plastic strains for CIU tests have not been reported in the figure.

Because of the high scatter of the yield points, neither of the two criteria seems to be suitable. In any case, the modified Cam-Clay yield curve offers a better

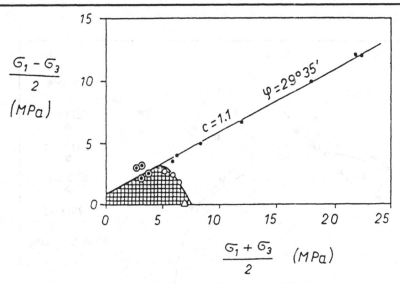

Fig. 12 - Neapolitan Yellow Tuff: rock-like and soil-like regions (after Pellegrino [8])

interpolation of the yield points and agrees with the direction of the plastic strain vectors. These vectors are supposed to be normal to the yield surface, when associated flow rule models are considered.

The scatter of the yield points can be attributed partly to the different values of initial porosity and partly to a different degree of bonding, and other structural features. As shown in Fig.4a, the isotropic yield stress significantly increases for decreasing initial porosity. Fig. 2 shows a similar influence of porosity on uniaxial compressive strength.

According to a normalisation procedure suggested by Critical State Mechanics with reference to soils, the influence of initial porosity could be eliminated dividing the yield stresses by a reference pressure p'o. This pressure is the mean effective stress corresponding to the intersection of the recompression line, relative to the initial porosity of the specimen, with the isotropic compression line [47].

Aversa et al. [22] applied this kind of normalisation to the yield points obtained on the TF tuff (see Fig. 14). The scatter of the data is reduced, but is still significant, as usually happens when the mechanical parameters of tuff are correlated with porosity. The yield curve reported in the figure has been drawn freehand.

A study of the effects of destructuration on the mechanical behaviour of the TF tuff during triaxial tests is in progress. Preliminary results indicate that the tuff specimens, previously subjected to isotropic compression with a maximum stress, approximately twice the isotropic yield stress, show a significant reduction in strength and increase in deformability.

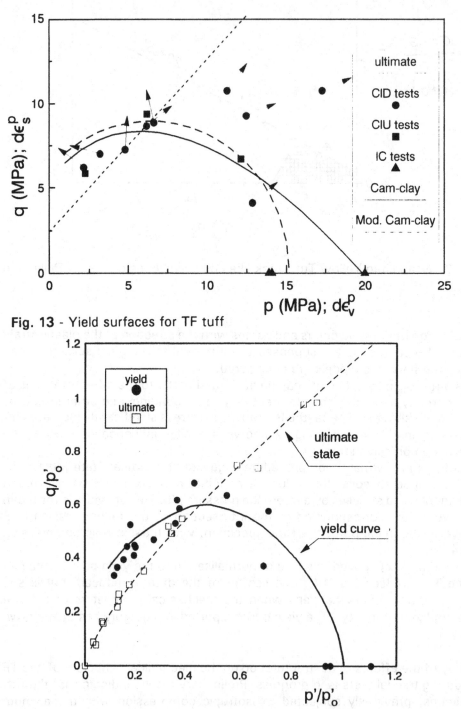

Fig. 13 - Yield surfaces for TF tuff

Fig. 14 - Yield surface for TF tuff in a normalised plane

5. Non-linearity in a boundary value problem

Evidence of the non-linear behaviour of tuff is given in plate loading tests.
Pellegrino [10] performed many tests in situ on plates with diameters of 30 and
50 cm following an incremental loading procedure. The tuff tested is present in
the Vomero hill in which a tunnel for the "Tangenziale" ringroad was driven.
This tuff is similar to the T1 tuff. The envelope of the results of loading tests
performed on the smaller (30 cm) plate is reported in Fig. 15.
A plate loading test was performed in the laboratory on the Fine-Grained Tuff
(TF) using a small plate with a 2.8-cm diameter (see Fig.16) at a constant rate
of displacement.
In both experimentations, the load - settlement curves present an initial linear
trend up to a stress approximately twice the uniaxial compressive strength,
followed by a non-linear hardening behaviour.

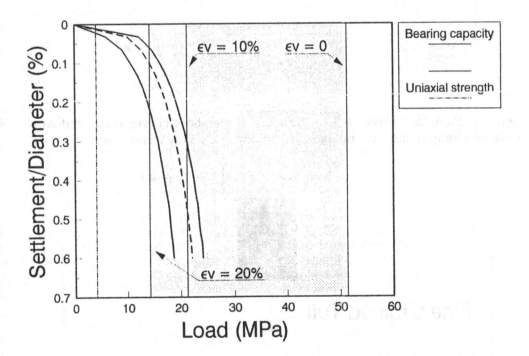

Fig. 15 - Plate load tests on Neapolitan Yellow Tuff: load - settlement envelope
compared with uniaxial strength and bearing capacity determined using Vesic
theory

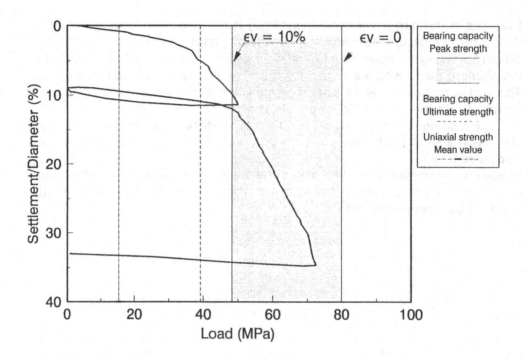

Fig. 16 - Plate load tests on TF tuff: load - settlements curves compared with uniaxial strength and bearing capacity determined using Vesic theory

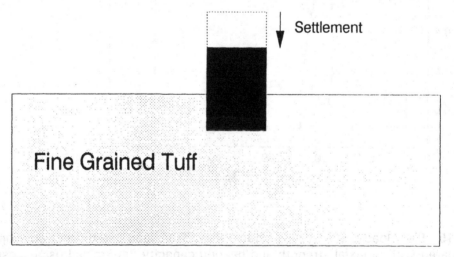

Fig. 17 - Plate load tests on TF tuff: sketch of the cross section at the end of the test

After the linear trend the plate begins to punch into the tuff mass. Only radial cracks are observed on the tuff surface. A bearing capacity is not clearly evidenced (see Fig.15 and 16).
A similar trend was observed for collapsible loess by Chercasov [48].
Settlements can be attributed to the progressive destructuration of a volume of tuff located immediately below the plate whose extension increases with the load. A sketch of a cross section of the Tf tuff at the end of the test is shown in Fig. 17.

An attempt was made to evaluate bearing using the Vesic theory [49].
The rigidity index in our case may be considered equal to $Ir = G/c$ where G is the shear modulus and c is the cohesion of the tuff. This index proves to be very high.
Taking into account the plastic volumetric strain ϵ_v developed during destructuration, the reduced rigidity index $Irr = Ir/(1 + Ir \, \epsilon_v)$ may be calculated. It assumes a lower value that produces a significant decrease in the compressibility factor and in the bearing capacity. This is in accordance with the punching behaviour clearly shown in Fig. 17.
The theory does not allow to predict the load settlement curve and a more complex analysis must be carried out using an appropriate stress - strain relation.

6. Time-dependent behaviour

6.1 Creep tests

The experimentation reported in this work was carried out at the end of the 1970s, but its results were never published. Recently laboratory experimentation was started again.
The tests have been limited to simple stress conditions, i. e. uniaxial compression and oedometric tests.
Constant stress has been maintained with simple dead load systems. Oedometers loading systems have been modified for this purpose. Available loading systems are linked with load amplifications of 1/15 (low pressure) and 1/47 (high pressure).
Both T1 and TF tuffs have been submitted to creep tests. Specimens have had a cylindrical form with a diameter equal to 10 cm for T1 and 5 cm for TF.
At the beginning of the experimentation conducted on the T1 tuff significant parassitic effects of the fluctuation of room humidity were noticed (see Fig.18).
This observation focused attention on the deformation of the rock induced by humidity variations. For the first time it was pointed out by Evangelista [11] that the volcanic tuff suffers elongation if its water content increases.
This behaviour is explained by the hydration of the zeolites present in the mass.

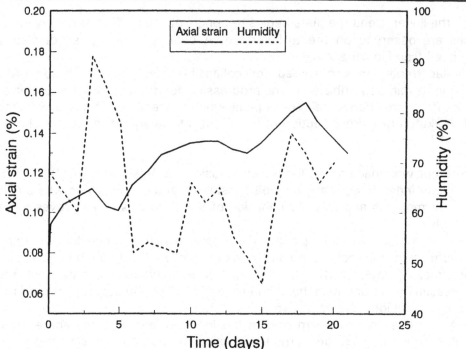

Fig. 18 - Parasitic effect of variation of atmospheric humidity during a creep tests on T1 tuff

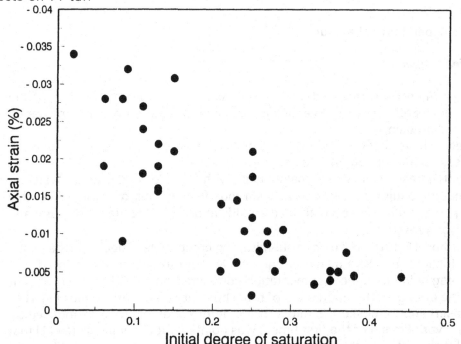

Fig. 19 - Axial strains induced by saturation of specimens of T1 tuff vs. initial degree of saturation

Table 2 - Summary of creep tests

Tuff	Specimen	Experimental conditions	Measured deformations	Number of loading steps	Maximum axial stress (MPa)	Assembly time (days)
T1	A11	room	local	10	3,78	100
T1	A70	room	local	9	3,00	255
T1	A83	in water at constant temperature	on the basis	6	2,80	22
T1	B3	room	local	9	3,00	254
T1	B15	room	local	8	3,32	32
T1	B39	room	local	13	3,48	29
T1	B44	room	local	10	2,82	41
T1	C52	room	local	4	1,11	256
T1	C88	room	local	6	1,90	253
T1	D8	in water at constant temperature	on the basis	8	3,21	34
T1	D39	room	local	10	4,49	239
T1	D59	room	local	6	2,00	11
T1	D88	constant water content	local	6	1,96	297
T1	E41	room	local	8	2,39	1
T1	E80	constant water content	local	6	1,96	291
TF	26	in water	on the basis	4	7,30	37
TF	32	in water	on the basis	4	7,72	116
TF	39	in water	on the basis	4	8,20	407

For each specimen the duration and the stress increment of each step has been arbitrarily selected according to the observed behaviour of the specimen itself.

It has recently been shown that tuffs containing different zeolitic minerals have different elongation [50].

The Neapolitan Yellow tuff is very sensitive to this phenomenon. In Fig. 19 the elongations undergone by the specimens of T1 tuff subjected to an axial stress of 1.5 MPa when submerged with water are plotted vs. their initial degree of saturation. These deformations, measured one day after the saturation of the specimen, are the algebraic sum of dilation produced by an increase in the water content, plus the creep contraction due to the load. The effect of humidity is considerable if we point out that axial strains at failure in uniaxial compression test are equal to a few tenths of %.

The capability of the T1 tuff to absorb water from the atmosphere is shown in Fig. 20. A specimen previously dried at 60 °C and then cooled in a stove is put

into a climatized room at 90% of relative humidity. Elongation of the specimen
is then measured. The process of water absorption and tuff elongation occurs
over a short time.

The parassitic effect of temperature is less important. Studies conducted on the
thermal behaviour of tuff have shown that the thermal expansion coefficient is
low as for other rocks and is approximately equal to 1E-05 1/°C [51, 52].

Some creep tests were initially conducted at room temperature and humidity,
while others were conducted on specimens submerged in water. Additional tests
were conducted in natural conditions after enveloping the specimen with a film
of plastic material and immerging it in latex, thus preventing variation in water
content.

Loads were applied by steps. Axial deformations were measured by mechanical
dial gauges. Local deformations were monitored by means of devices placed
along two opposite generating lines on the lateral surface of the specimens or
at their bases.

In Table 2 uniaxial tests performed on both types of tuff are summarised.

Figures 21 and 22 show typical results obtained from uniaxial creep tests
performed on the tuff specimens.

An oedometric compressive test was conducted in a high pressure oedometer
in room conditions for some days.

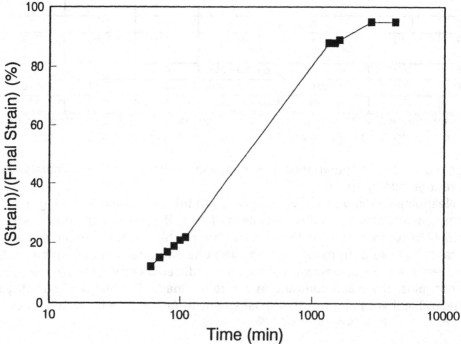

Fig. 20 - Elongations experienced by a dried specimen of T1 tuff put into a
climatised room with 90% of relative humidity

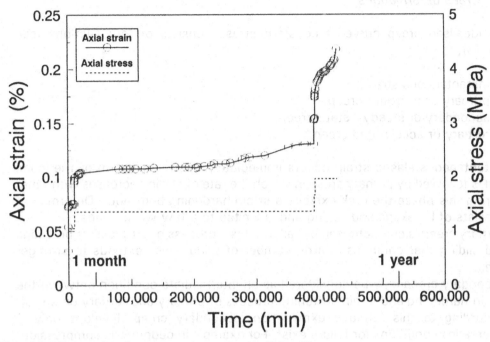

Fig. 21 - Example of creep test results on T1 tuff

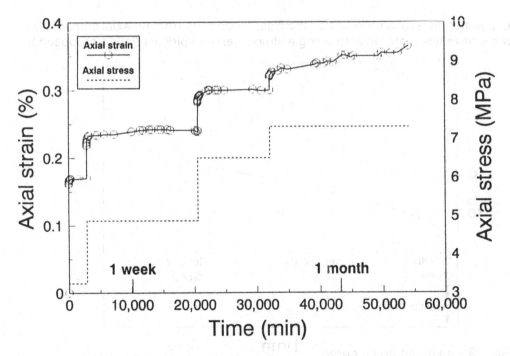

Fig. 22 - Example of creep test results on TF tuff

6.2 Creep deformations

An idealised creep curve at constant stress consists of four sections (see Fig.23):

1) instantaneous strain,
2) primary or transient creep,
3) secondary or steady - state creep,
4) tertiary or accelerated creep.

Instantaneous elastic strain occurs immediately upon application of the loads and is followed by primary creep in which the rate of strain decreases with time. During this phase the rock exhibits a strain-hardening behaviour. Deformation consists of breakage and sliding and is limited to a few weak bonds.
Tertiary creep is characterised by failure. It is a process of progressive breakage and sliding that concerns a large number of bonds and extends to stronger ones.
Secondary creep may be considered as an intermediate condition between the strain-hardening and strain-softening process of primary and tertiary creep [5]. According to this intuitive explanation, secondary creep develops only if kinematics conditions for failure exist. For example in oedometric compression, only primary creep occurs.

Experimental results obtained on the Neapolitan tuffs from uniaxial compression tests have been interpolated using a simple semi-empirical relation proposed by

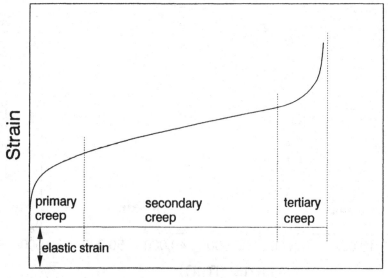

Fig. 23 - Idealised creep curve

many authors and reported by Lama and Vutukuri [53]:

$$\varepsilon = A + B \log\left(\frac{t}{t_1} + 1\right) + C t \tag{5}$$

This correlation has been preferred to simple polynomial functions because it permits to distinguish between elastic strains (parameter A), primary creep (B log $(t/t_1 + 1)$) and secondary creep (C t). The time t_1 represents a reference time (equal to 1 minute in this paper).

Equation (5) is unable to interpret tertiary creep. This phase has not been investigated in appreciable detail.

Constants A, B and C of Eq. (5) depend upon the state of stress. In the case of a single step test, A, B and C are related to the applied stress. In a stepwise increasing stress process, constants A, B and C must be related to the current stress reached σ_v and to the amplitude of the stress increase $\Delta\sigma_v$.

In Fig. 24 the interpolation of the results a single step procedure is reported as an example.

It is evident that the longer the duration of the steps, the more accurate the distinction between primary and secondary creep (respectively expressed by

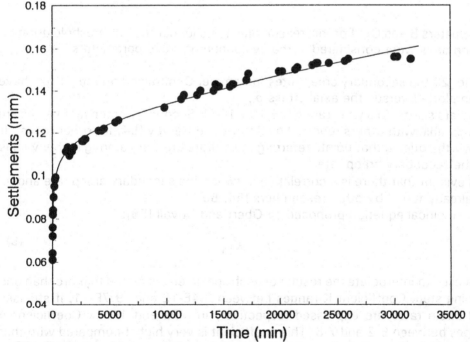

Fig. 24 - Example of interpolation of time - settlement data for a single step of loading on T1 tuff

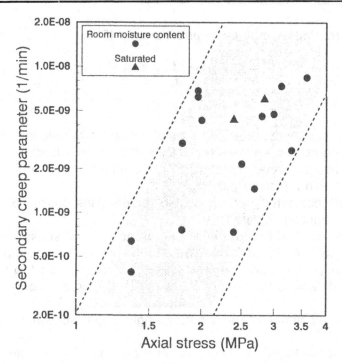

Fig. 25 - Secondary creep rates obtained from tests on T1 tuff vs. axial stress

parameters B and C). For this reason only tests longer than a threshold duration
of ten days were considered in the evaluation of creep parameters.

In Fig. 25 the secondary creep rates (parameter C) obtained on the T1 tuff have
been plotted versus the axial stress σ_v.
The data scatter in a wide range (1E-10 ÷ 1E-8). Secondary creep rate values are
comparable with others reported by Afrouz and Harvey [54] for other rocks. It
is worth noticing that points referring to saturated tuff are among higher values
of the secondary creep rate.
It is evident that there is a correlation between the secondary creep rate and σ_v
as already found by other researchers [54, 55].
The empirical equation proposed by Obert and Duvall [55]:

$$\dot{\varepsilon} = K\sigma_v^n \qquad (6)$$

was used to interpolate the results of each specimen subjected to more than one
loading step. Coefficient K ranges between 3.4E-12 and 9.7E-11, if stresses
and strain rates are expressed respectively in MPa and 1/min. Coefficient n
ranges between 6.2 and 7.8. This coefficient is very high if compared with that
obtained by other researchers [55] on different hard rocks, for which it assumes
values of between 1.0 and 3.3. The dashed lines in Fig. 25, characterised by

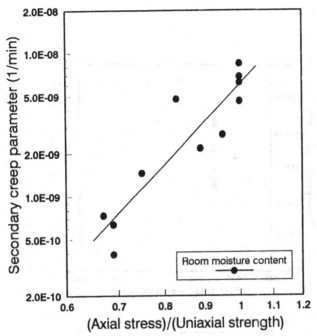

Fig. 26 - Secondary creep rates obtained from tests on T1 tuff vs. axial stress level

a slope equal to the average value of n, delimits the range of experimental results.

The wide range of K parameters, graphically expressed by the distance between the dashed lines, is due to the different strength of the tested specimens. In order to normalise the data, by eliminating the influence of strength, the following correlation has been proposed:

$$\dot{\varepsilon} = K_1 \left(\frac{\sigma_v}{\sigma_r}\right)^m \tag{7}$$

in which σ_r represents the uniaxial strength of the material measured at the end of creep tests conducted up to failure.

In Fig. 26 experimental results have been plotted in a C vs. σ_v/σ_r diagram. The best fitting function is defined by a coefficient K_1 and m respectively equal to 6.0E-9 and 5.8, if stresses and strain rates are expressed in MPa and 1/min.

The values of the primary creep parameter B for tests longer than ten days are reported in Figs. 27a and 27b versus vertical axial stress and vertical axial stress increments respectively.

The data in both figures are extremely scattered. The primary creep parameter B ranges between 1E-7 and 1E-4. Data concerning saturated specimens are

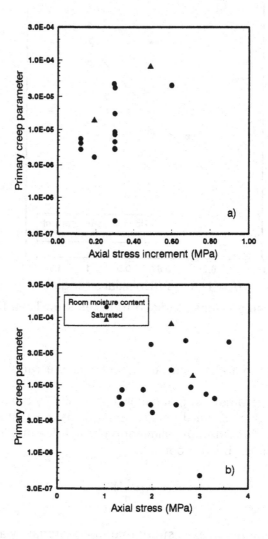

Fig. 27 - Primary creep parameter B obtained from tests on T1 tuff: a) B vs. axial stress increment; b) B vs. axial stress

located in the upper part of the diagrams.

Besides the scatter, it seems that the primary creep parameter increases with the axial stress increment. No correlation seems to exist between this parameter and the vertical stress.

In order to highlight the influence of stress levels and to hide the influence of stress increases on primary creep, the ratio between the primary creep parameter B and the elastic strain versus the axial stress is given in Fig. 28.   It is evident that this ratio increases with the axial stress. These observations show that viscous effects increase, expectedly, with stress level.

From the comparison between Fig.25 and Fig.28, it appears that the influence of the stress level on secondary creep is much more evident than that on primary creep.

In order to stress the importance of primary creep, the ratio $\epsilon_{(10\ days)}/\epsilon_{(1\ min)}$ is plotted vs. the axial stress in Fig. 29. An exponential regression line has been used to interpolate experimental data. The influence of stress level on primary creep is clearly shown in the figure.

It is interesting to notice (see Fig. 29) that the T1 tuff seems to show a threshold of 0.8 MPa. Below this value creep is practically absent. This value corresponds approximately to 1/3 - 1/4 of the uniaxial strength of the material.

As already mentioned at the beginning of the paragraph, no remarks have been made on the pattern of tertiary creep.

The previous remarks pertain to the uniaxial compression tests. Along this stress-path the behaviour is approximately "linear" up to failure, as stressed in

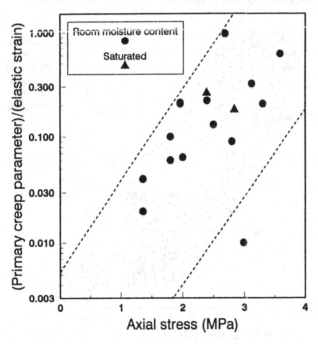

Fig. 28 - Ratio between primary creep parameter B and elastic strain obtained from tests on T1 tuff vs. axial strain

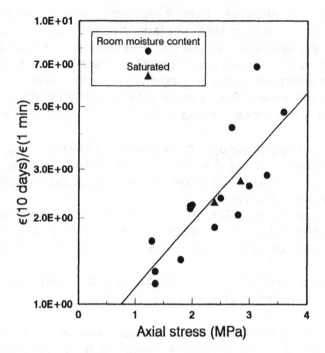

Fig. 29 - Ratio between ϵ(10 days) and ϵ(1 min) vs. axial strain

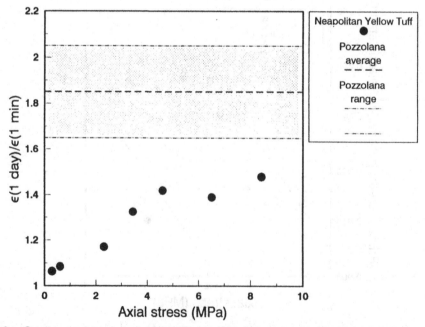

Fig. 30 - Oedometric test on T1 tuff: ratio between ϵ(1 day) and ϵ(1 min) vs. axial stress; comparison with pyroclastic soil range

paragraph 3.2. In triaxial conditions, for higher values of the mean effective stress the behaviour becomes non-linear and essentially plastic. In this stress range an increase in creep deformations is to be expected. In order to confirm this idea, a simple incremental loading oedometric test was performed on the San Rocco tuff (T1).

In oedometric conditions, because of the kinematic restraints, only primary creep occurs.

In Fig. 30 the ratio between one-day and one-minute strains is related to the applied axial stress. The creep strains quickly increase with stress up to 5 MPa, which represents the yield stress in oedometric conditions. For higher stresses the slope of the curve becomes flatter. The range of this ratio obtained by Pellegrino [6] on "pozzolana" is reported in Fig. 30. for purposes of comparison. It is evident that the ratio $\epsilon_{(1\ day)}/\epsilon_{(1\ min)}$ of tuff seems to tend to lower pozzolana values. It is possible to infer that for stress levels far from failure, but outside the elastic domain, creep deformations become important and assume values that are very close to those of soils.

6.3 Long time strength

Experiments with many materials like polymers, metals and natural soils and rocks have clearly shown that strength is a time-dependent feature.

The stress that a material may sustain without failure depends upon the time the load acts just as the weight that a man is capable of sustaining depends upon his endurance.

An important implication of this behaviour is the evaluation of the maximum stress at which no failure occurs, no matter how long the load is applied.

This strength has been termed differently as "fundamental strength", "true strength", "time safe stress", "sustained load strength" and "long term strength".

Studies conducted on soils and rocks have shown that the "long term strength" is very close to ultimate strength [5, 29, 56, 57] (see Fig. 31).

The loss of strength with time depends upon the drop in cohesion. In exploring Zabraslav sand, Feda [5] shows that the friction angle is time-independent, in agreement with Kenney's [58] results. Cohesion decreases with time to a small but non-zero value. Such a decrease is due to strains induced in the creep process.

Lama and Vutukuri [53] reported different techniques, proposed by various researchers, intended to determine the "long term strength" of rocks.

The evaluation of this strength for the Neapolitan volcanic tuff was performed on the basis of the above-mentioned creep tests (see Tab. 2) and of uniaxial compression tests performed at different strain rates.

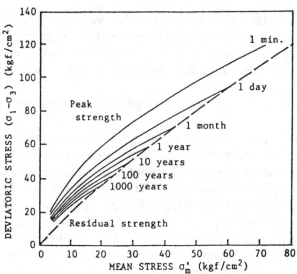

Fig. 31 - Long term strength of a Japanese volcanic tuff

Monkman and Grant [59] showed that the time of creep failure of different materials is related to the minimum creep rate by means of the power function:

$$t_f = D\dot{\varepsilon}^r \qquad (8)$$

In this equation the coefficient r varies from 0.77 to 0.93 for a large variety of materials
Saito and Uezawa [60] specialised this relation for soils, obtaining the following relation:

$$\log(t_f) = 1.334 - 0.916\log(\dot{\varepsilon}) \pm 0.59 \qquad (9)$$

Feda [5] reported a similar correlation for other kinds of soils and weak rocks. Adachi and Takase [56] proposed the following correlation for a Japanese volcanic tuff:

$$t_f = 0.00132\,\dot{\varepsilon}^{-1.0} \qquad (10)$$

Only a few creep tests performed on the Neapolitan tuffs reached failure conditions. The data reported in the creep failure time vs. minimum creep rate diagram (Fig.32) was obtained from these tests. The data relative to the non-saturated San Rocco tuff (T1) has been interpolated with a power function as proposed by previously mentioned researchers. The best fitting curve has been drawn in the figure and has the following expression:
This curve falls outside the range identified for other weak rocks [5] and it is

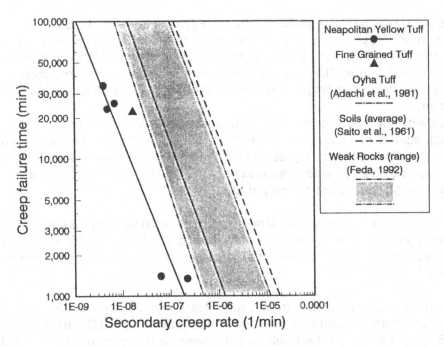

Fig. 32 - Relationship between creep failure time and secondary creep rate

$$t_f = 0.0009453 \, \dot{\varepsilon}^{-0.896} \tag{11}$$

below the curves determined by Saito and Uezawa [60] for soils and Adachi and Takase [56] for another tuff. For a fixed minimum strain rate the Neapolitan volcanic tuffs are characterised by a shorter time of creep failure.

According to the strain rate method for evaluating long time strength proposed by Sangha and Dhir [61], the non-saturated T1 tuff and the non-saturated TF tuff were tested at different strain rates. In Fig. 33 uniaxial compressive strengths for the TF tuff are plotted versus strain rates. Compressive strength decreases as strain rate decreases.

In Fig.34 data obtained from creep failure tests and strain controlled tests are summarised for T1 tuffs. A time was assigned to each strain rate according to the following procedure. The strain rate was confused with minimum creep rate according to Ohtzuki et al. [29]. Time was inferred from a correlation (11) determined for the Neapolitan tuffs.
On the basis of the failure points, a statistic correlation was obtained which is indicated by the solid line in the figure:
where σ_1 = 5.05 MPa is the uniaxial compressive strength at 1 minute. The

coefficient b is equal to 0.55 MPa. The ratio β/σ_1 is equal to 10.89%.

In order to increase the number of data corresponding to longer times, other points representative of conditions before failure in creep tests have also been reported in the figure. They indicate the time during which a stress greater or equal to that corresponding to the y value of the point has been applied to the specimen.

It is evident that these points should be located below the failure line.

The high scatter of creep data justifies the presence of some of these points above the solid line which represents an average of the failure conditions. An upper bound curve has also been drawn in the figure. The lower bound curve passes through the lowest failure points.

Experimental data obtained from strain controlled uniaxial compression tests on non-saturated fine-grained tuffs (TF) have been plotted (see Fig.35) versus time calculated according to the correlation between time to failure and strain rate obtained for the T1 tuff.

Experimental data obtained from creep tests and strain control uniaxial compression tests performed on saturated fine-grained tuffs (TF) are shown in Fig. 36. The meaning of the symbols is the same as those used for the T1 tuff. Data of creep tests are very few and therefore the statistical correlation reported

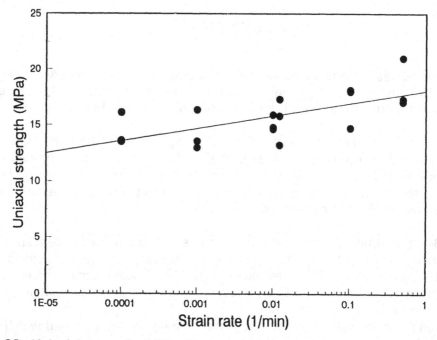

Fig. 33 - Uniaxial strength of TF tuff obtained by means of tests performed with different axial strain rates

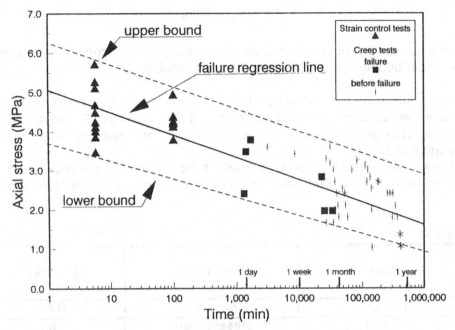

Fig. 34 - Dependence of time to failure to stress for T1 tuff

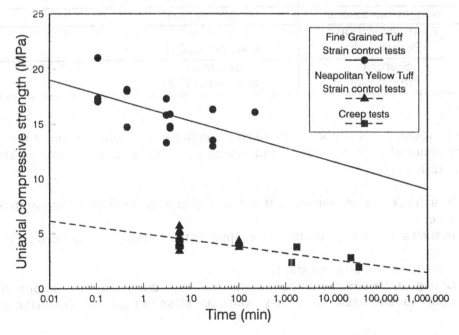

Fig. 35 - Dependence of time to failure to stress for non-saturated TF tuff

$$\sigma = \sigma_1 - \beta \log(t) \qquad\qquad (12)$$

in the figure may be considered very approximate. The correlation relative to the upper bound curve appears to be more reliable.

A comparison of the influence of time on the compressive strength of the volcanic Neapolitan tuff and of other rocks is given in Table. 3. The ratio β/σ_1 of equation(12) given in this table was obtained by processing data found in the literature [53].
As one may notice Neapolitan volcanic tuffs fall among the rocks with the largest values of the β/σ_1 ratio which come close to the values of Potash Salt.

Table 3

Rock	Reference	β/σ_1 (%)
T1 Tuff	present work	10.89
TF tuff	present work	7.60
Saturated TF tuff	present work	5.83-8.19
Westerly Granite	Wawersik [62]	6.62
Nugget Sandstone	Wawersik [62]	5.49
Verkhnekamskaya Potash Salt	Stavrogin and Lodus [63]	9.89
Starobin Potash Salt	Stavrogin and Lodus [63]	11.39

In conclusion it is important to make a comprehensive comparison of the time behaviour of tuffs with respect to that of other rocks. It is possible to stress the following points:

- In a "rock like" dominion, tuff exhibits, a significant loss of strength with time;
- In such a dominion, the strain rate, appears to be strongly influenced by stress level;
- Time to failure appears short;
- Owing to the destructuration, in the "soil like" dominion, time deformations seem to reach relevant values which are close to those of pyroclastic soils.

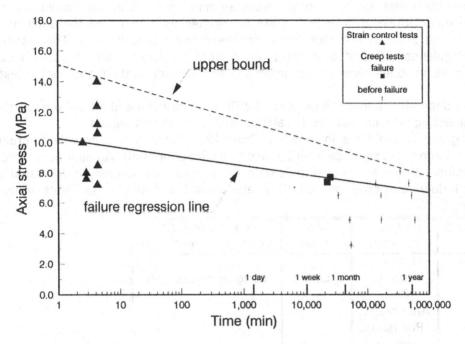

Fig. 36 - Dependence of time to failure to stress for saturated TF tuff

7. Conclusions

It seems interesting to make some conclusive remarks on the technical role of non-linearity and creep of tuff.
It is necessary to differentiate between technical problems characterised by low mean stresses and those characterised by high mean stresses (see table 4).

Table 4

Group	Technical problem	Mean stress	Factor affecting limit state
A	Tunnel, Quarry, Cliff, Anchorage, Pile (shaft resistance)	Low	Failure
B	Shallow foundation Pile (tip resistance)	High	Yield

It is evident that non-linearity assumes an important rule in the problems with high mean stresses. The limit state is reached as a result of the yield of a contained zone of the rock in which the mean stress is high. This aspect distinguishes tuff and other soft rocks from hard rocks and soils. A clear example has been given in paragraph 5 with reference to the plate loading test.

Time-dependent behaviour assumes a different role in dependance of the various engineering problems at hand. Table 5 summarises this aspect.
For group A problems in which a "rock-like" behaviour is prevalent, creep deformations are significant only close to failure conditions. An adequate factor of safety, for instance equal to 3, seems practically vanish creep deformations. The influence of time-to-strength is relevant at low mean stress levels where

Table 5

Group	Technical problem	Mean stress	Feature of time-dependent behaviour
A	Tunnel, Quarry, Cliff, Anchorage, Pile (shaft resistance)	Low	Long-term strength
B	Shallow foundation Pile (point resistance)	High	Time deformations

the role of bonds (cohesion) is important. In this field an important loss of strength has been observed with time. In particular the influence of time induces a contraction of the elastic domain, a reduction in cohesion and an increase in the friction angle that leads to an ultimate value.
For the second group of problems, even if the safety factor is high, creep deformations are significant with respect to immediate deformations. Fortunately, from a technical point of view, the problem is of a relative importance owing to the small size of the deformations.
The bearing capacity is poorly influenced by the loss of cohesion which indeed is compensated by the increase in the friction angle. The contraction of the elastic domain may induce plastic deformations.

In conclusion it may be stressed that the time-dependent behaviour plays a main role in Group A problems (Table 5). This remark is validated by certain events which have taken place in Naples [64]. Many tunnels excavated in the Neapolitan tuff under the Posillipo hill presented major damage a few years after construction. The "Quattro Giornate" tunnel opened in 1884 was closed in 1890 for considerable damage. The railway tunnel, positively tested in 1917, showed some cracks in 1922 and was closed for restoration works until 1925. The "Laziale" tunnel completed in 1926 suffered damages in 1932. The "Circumflegrea" Railway tunnel, excavated in the 1954 under the Vomero hill, had to be restored around 1967.

In the past, tuff used to be quarried by means of underground excavations, in order to be used as a building material and with time. The quarries, situated just outside the borders of the existing city, would be abandoned, but no action would be taken to ensure their stability. When the town expanded, buildings were erected on the soil overlying such underground quarries, thus creating serious stability problems. For economic reasons, excavations in the quarries had been so extensive as to induce high stresses in the pillars and in the vaults, thus frequently causing the development of dense networks of large fractures, and falls of large blocks both from the vaults and from the pillars. Also in this case failure phenomena sometimes occur long after the excavation has been completed.

One of these quarries excavated between the seventeenth and the eighteenth centuries was subjected to an accurate investigation since a water pumping station was to be installed.

The mean values of vertical stresses in the pillars were calculated and specimens were extracted from pillars and tested under uniaxial compression [12].

Fig. 37 shows the mean value of stress and the mean uniaxial strength for each

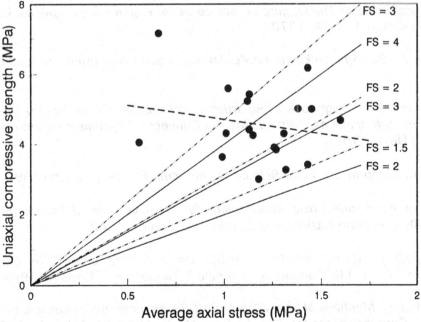

Fig. 37 - Mean values of uniaxial compressive strength and vertical stress relative to different pillars of the Reichlin quarry excavated in Neapolitan Yellow Tuff

pillar. Equal factors of safety lines have also been drawn in the figure. The factors of safety, indicated with the solid line, are calculated according to the uniaxial strength reported in the figure. The factors of safety indicated with the dashed and dotted lines are calculated taking into account the reduction in strength due to size and degree of saturation effects [18]. When referring to the latter lines, the factors of safety assume pretty small values, which justify the presence of creep phenomena in the pillars and therefore a stress redistribution in both pillars and vaults. This remark is confirmed by the presence of cracks in the pillars and by the falls of large blocks from the vaults.

The excavation of this quarry and its following life can be considered as a long-term creep test (loading time is approximately 350 years), from which it is possible to obtain the influence of applied stresses on the strength of the material. Beside the high scatter of the data, the linear regression (dashed line) indicates that strength decreases as the mean vertical stress increases. This observation seems to indicate that also low stresses, if applied for a long time, can induce a reduction in strength.

References

1. AGI: *Raccomandazioni sulle esecuzioni delle indagini geotecniche.* Associazione Geotecnica Italiana, Roma 1977.

2. Geological Society: *The logging of rock cores for engineering purposes.* Q.J. Engng. Geol. 3, n.1: 1-24, 1970.

3. Ogawa Y.: *Geology Ishikawa, no.40.* Ishikawa Soil Corporated Association, 1986.

4. Evangelista A., Pellegrino A.: - *Ripartizione delle tensioni tra scheletro solido ed acqua nel tufo giallo napoletano*. Atti del Convegno "Bradisismo e fenomeni connessi", Napoli, 1986.

5. Feda J.: *Creep of Soils and Related Phenomena*, Elsevier, London 1992.

6. Pellegrino A.: *Proprietà fisico-meccaniche dei terreni vulcanici del napoletano.* Atti del VIII Convegno Nazionale di Geotecnica, Cagliari 1967.

7. Pellegrino A.: *Compressibilità e resistenza a rottura del Tufo Giallo Napoletano.* Atti del IX Convegno Nazionale di Geotecnica - Genova, 1968.

8. Pellegrino A.: *Mechanical behaviour of Soft Rocks under high stresses.* Proc. of the 2nd. Congress of the Int. Soc. for Rock Mech., Vol.3-25, Belgrade, 1970.

9. Pellegrino A.: *Le rocce lapidee tenere.* Ricerche in corso: risultati e linee di sviluppo del Gruppo di ricerche Terreni e Strutture del CNR, Napoli, 1970.

10. Pellegrino A.: *Surface footings on Soft Rocks.* Proc. of the Third Congress of the International Society of rock Mechanics, Denver 1974.

11. Evangelista A.: *Influenza del contenuto d'acqua sul comportamento del tufo giallo napoletano.* Atti del XIV Convegno Nazionale di Geotecnica, Firenze 1980.

12. Evangelista A., Lapegna U., Pellegrino A.: *Problemi geotecnici nella città di Napoli per la presenza di cavità nella formazione del Tufo.* Atti del XIV Convegno Nazionale di Geotecnica, Firenze 1980.

13. Paparo Filomarino M., Pellegrino A. *L'effetto dimensione nel Tufo Giallo Napoletano.* Atti del XIV Convegno Nazionale di Geotecnica, Firenze 1980.

14. Aurisicchio S., Evangelista A., Masi P.: *Il Tufo Giallo Napole-tano: permeabilità ed impregnabilità con monomeri acrilici.* Rivista Italiana di Geotecnica (1956) Anno XVI N.1.

15. Aurisicchio S., Masi P., Nicolais L., Evangelista A., Pellegrino A.: *Polymer impregnation of Tuff.* Polymer Composites (1982), July, Vol.3 , N.3.

16. Rippa F., Vinale F.: *Structure and Mechanical Behaviour of a Volcanic Tuff.* Proc. of V Congress of ISRM, Melbourne 1983.

17. Aurisicchio S., Evangelista A., Nicolais L.: *Comportamento meccanico del tufo trattato con polimeri.* Rivista Italiana di Geotecnica (1985), Anno XIX, n. 2.

18. Evangelista A., Pellegrino A.: *Caratteristiche geotecniche di alcune rocce tenere italiane.* Atti del terzo ciclo di Conferenze di Meccanica e Ingegneria delle Rocce MIR90, Torino 26 - 29 Novembre 1990.

19. Aversa S., Evangelista A., Ramondini M.: *Influenza della struttura sulla compressibilità di un tufo vulcanico.* Atti della Riunione Annuale dei Ricercatori del G.N.C.S.I.G. del C.N.R., 157 -160, Roma 1990.

20. Aversa S., Evangelista A., Ramondini M.: *Snervamento e resitenza a rottura di un tufo a grana fine.* Atti del II Convegno dei Ricercatori del G.N.C.S.I.G. del C.N.R., Vol. 1: 3 - 22, Ravello 1991.

21. Aversa, S.: *Mechanical behaviour of soft rocks: some remarks.* Proc. of the Workshop on "Experimental characterization and modelling of soils and soft

rocks", 191-223, Napoli 1991.

22. Aversa S., Evangelista A., Leroueil S., Picarelli L.: *Some aspects of the mechanical behaviour of "structured" soils and soft rocks.* Proc. of International Symposium on "Geotechnical Engineering of Hard Soils and Soft Rocks", vol.1: 359-366, Athens 1993.

23. Leroueil, S., Vaughan, P.R.: *The general and congruent effects of the structure in natural soils and weak rocks.* Géotechnique (1990), Vol. 40 (3): 467-488.

24. Conrad H.: *The role of grain boundaries in creep and stress rupture.* In Dorn J. E. (Ed.) Mechanical Behaviour of materials at elevated temperatures: 218 - 269, Mc. Graw Hill New York 1961.

25. Lubahn J. D.: *Deformation Phenomena.* In Dorn J. E. (Ed.) Mechanical Behaviour of materials at elevated temperatures: 319-392, Mc. Graw Hill New York 1961.

26. Quingley R.M., Thompson C.D.: *The fabric of anisotropically consolidated sensitive marine clay.* Can. Geot. J. (1966), Vol. 3, n.2: 61-73.

27. Yoshinaka R., Yamabe T. - *Deformation behaviour of soft rocks.* Proc. of the Int. Symp. on Weak Rocks: 87 - 92, Tokyo 1981.

28. Travi B.: *Comportamento meccanico dei terreni strutturati e delle rocce lapidee tenere.* Tesi di Laurea (unpublished), Napoli 1993.

29. Ohtsuki H., Nishi K., Okamoto T., Tanaka S.: *Time-dependent characteristics of strength and deformation of a mudstone.* Proc. Int. Symp. on Weak Rocks, Vol. 1: 107 - 112, Tokyo 1981.

30. Elliot G.M., Brown E.T.: *Yield of a soft, high porosity rock.* Geotechnique (1985), Vol.35, No.4: 413 - 423.

31. Elliot G.M., Brown E.T.: *Further developments of a plasticity approach to yield in porous rocks.* Int. J. Rock Mech. Min. Sci. & Geomech. Abstr. (1986), Vol. 23, No. 2: 151 - 156.

32. Maekawa H.: *Mechanical Properties of soft rocks and indurated soils.* Report of the technical Committee on Soft Rocks and Indurated Soils, Rio de Janeiro 1989.

33. Lefebvre G.: *Contribution à l'etude de stabilité des pentes dans les argiles*

cimentées. Ph.D. Thesis, Laval University, Quebec 1978.

34. Yoshinaka R., Yamabe T.: *A strength criterion of rocks and rock masses.* Proc. of the Int. Symp. on Weak Rock: 613 - 618, Tokyo 1981.

35. Price A.M., Farmer I.W.: *Application of yield models to rock.* Int. J. Rock Mech. Min. Sci. & Geomech. Abstr. (1979), Vol. 16: 157 - 159.

36. Price A.M., Farmer I.W.: *The Hvorslev surface in rock deformation.* Int. J. Rock Mech. Min. Sci. & Geomech. Abstr. (1981), Vol. 18: 229 - 234.

37. Akai K., Adachi T., Nishi K.: *Mechanical Properties of Soft Rocks.* Proc. of IX ICSMFE, Vol. 1: 7-10, Tokyo 1977.

38. Adachi T., Ogawa T., Hayashi M.: *Mechanical properties of soft rocks and rock mass.* Proc. of the 10th ICSMFE, Vol.1: 527 - 530, Stockholm 1981.

39. Clough, G.W, Sitar, N., Bachus, R.C., Rad, N.S.: *Cemented sands under static loads.* J. Geotech. Engrg. Div. ASCE (1981), Vol. 107 (6): 799-817.

40. Maccarini M.: *Laboratory studies of a weakly bonded artificial soil.* Ph.D. Thesis, Imperial College of Science and Technology, London (U.K.) 1987.

41. Lade P.V., Overton D.D.: *Cementation Effects in Frictional Materials.* Journal of Geotechnical Engineering ASCE (1989), Vol. 115, No. 10: 1373 -1387.

42. Gerogiannopoulos N.G., Brown E.T.: *The critical state concept applied to rock.* Int. J. Rock Mech. Min. Sci. & Geomech. Abstr. (1978), Vol. 15: 1 - 10.

43. Nova R. : *Discussione su: Snervamento e resistenza a rottura di un tufo a grana fine* di Aversa S., Evangelista A., Ramondini M. Atti del II Convegno dei Ricercatori del G.N.C.S.I.G. del C.N.R., Vol. 2, Ravello 1991.

44. Nova R.: *An extended Cam-Clay model for soft anisotropic rocks.* Computers and Geotechnics (1986), Vol.2: 69 - 88.

45. Esaki T., Kimura T., Nishida T.: *The Yield Function with End-Cap Based on Plasticity Theory.* Proc. of the Second Int. Conf. on Constitutive Laws for Engineering Materials, Vol.1: 387 - 394, Tucson - Arizona 1987.

46. Nova R.: *Deformabilità, resistenza meccanica, modelli costitutivi delle rocce tenere.* Atti del terzo ciclo di Conferenze di Meccanica e Ingegneria delle Rocce MIR90, Torino 1990.

47. Wood D.M.: *Soil Behaviour and Critical State Soil Mechanics.* Cambridge University Press 1990.

48. Chercasov I.I.: *The residual deformations caused by the deep penetration of a rigid plate into a brittle porous material.* Proc. of ICSMFE, Vol. 1. Oslo 1967.

49. Vesic S.A.: *Bearing Capacity of Shallow Foundations.* In: Winterkorn H.F. and Fang Y. (Eds.) "Foundation Engineering Handbook" :121-147, 1975.

50. Marino O., Macolo G., Cioffi R., Colantuono A., Dal Vecchio S.: *Tufi vulcanici: meccanismi di deterioramento chimico-fisico e tipologia di intervento.* L'Edilizia, n. 9: 523-535, 1991.1

51. Aversa S.: *Comportamento del Tufo Giallo Napoletano ad elevate temperature.* Tesi di Dottorato di Ricerca in Ingegneria Geotecnica, Università di Napoli 1989.

52. Aversa S., Evangelista A.: *Thermal expansion of Neapolitan Yellow Tuff.* Rock Mechanics and Rock Engineering, 1993 (in print).

53. Lama R.D., Vutukuri V.S.: *Handbook on Mechanical Properties of Rocks.* Trans Tech Publications, 1978.

54. Afrouz A., Harvey J.M.: *Rheology of rocks within the soft to medium strength range.* Int. J. Rock Mech. Min. Sci. & Geomech. Abstr. (1974), Vol. 11, n. 7: 281- 290.

55. Obert L., Duvall W.I.: *Rock Mechanics and the Design of Structures in Rocks.* Wiley, New York 1967.

56. Adachi T., Takase A.: *Prediction of long term strength of soft sedimentary rocks.* Proc. of the Int. Symp. on Weak Rock: 99 - 104, Tokyo 1981.

57. Okamoto: *Multistage creep tests analysis of mudstone.* Proc. of XI ICSMFE, Vol.2: 1027-1030, San Francisco 1985.

58. Kenney: *A Review of recent researches on strength and consolidation of soft sensitive clays.* Can. Geot. J.(1968), Vol. 5, N. 2: 97- 119.

59. Monkam F. C., Grant N.J.: *An empirical relation between rupture life and minimum creep rate in creep rupture tests.* ASTM Proc. (1956), Vol. 56: 593-605.

60. Saito M., Uezawa H.: *Failure of soils due to creep.* Proc. of V ICSMFE, Vol. 1: 315-318, Paris 1961.

61. Sangha C.M., Dhir R.K.: *Influence of time on the strength deformation, and fracture properties of a Lower Devonian Sanstone.* Int. J. Rock Mech. Min. Sci. (1972), Vol. 9, N.3: 343-354.

62. Wawersik W. K. *Time-dependent rock-behaviour in uniaxial compression.* XIV Symp. Rock. Mech.:85-106, University Park, Pennsylvania 1972.

63. Stavrogin A. N., Lodus E.V.: *Creep and time dependence of the strength in rocks.* Sov. Min. Sci. (1974), Vol. 10, N. 6: 653-658.

64. Ferrari P., Paolella G.: *Gallerie stradali e ferroviarie.* Atti dello VIII Convegno di Geotecnica, Cagliari 1967.

VISCOPLASTICITY OF GEOMATERIALS

N.D. Cristescu
University of Florida, Gainesville, FL, USA

ABSTRACT

First one describes the main rheological properties exhibited by geomaterials and how to choose the most appropriate constitutive equation in order to describe these dominant properties. One presents the procedure to be followed in order to determine all constitutive functions and parameters necessary to describe transient and stationary creep. Examples are given as well as a comparison of model predictions with the data. Damage and creep failure is described by an energetic criterium. Temperature influence on transient and stationary creep is shortly presented. Finally mining engineering examples are given as creep around vertical shaft or cavern and around horizontal tunnel, dilatancy and/or compressibility, failure and creep failure around underground excavations, closure of caverns, etc.

Course given at the International Centre for Mechanical Sciences (CISM), Udine, October 11-15, 1993.

1. RHEOLOGICAL PROPERTIES OF GEOMATERIALS AS EXHIBITED BY TESTS

In order to determine the main mechanical properties exhibited by geomaterials one performs several "diagnostic" tests just to reveal these properties. The main tests of this sort and the results obtained will be described schematic below.

1.1 Rate influence.

In all laboratories geomaterials are tested either in uniaxial or in triaxial tests. The specimen is either a cylinder, with the diameter to length ratio equal to ½, or a cubical specimen. No details concerning the experimental setup will be given here. We would like to describe only the results which are obtained. Let us consider first short-term uniaxial tests, where only one stress component is assumed to be nonzero. Generally the axial strain ε_1 and the circumferential (diameter) strain $\varepsilon_2=\varepsilon_3$ are measured. From $\varepsilon_v=\varepsilon_1+2\varepsilon_2$ follows also the volumetric strain. By convention compressive stresses are positive and strains describing a shortening are also positive.

Typical uniaxial stress-strain curves for schist are shown in Fig.1.1. Three loading rates shown were used. Thus with the increase of the loading rate, the whole stress-strain curve raises starting from the smallest values of the stress and strain. Fig.1.1

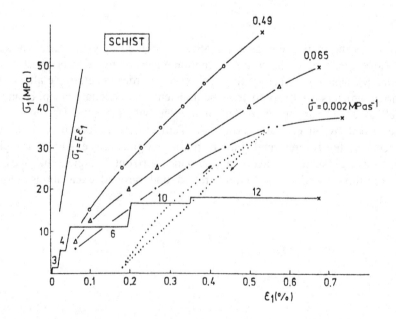

Fig.1.1   Stress-strain curves for schist for various loading rates, showing time influence on the whole stress-strain curves, including failure.

also shows the result of a creep test, the stress being held constant at various levels for 3 days, then for 4 days, etc., as shown (total 35 days). Let us point out that creep occurs starting from the smallest values of the loading stresses. Similar results have been reported for rock salt by HANSEN *et al.*[1984]. Since the stress-strain curves are loading rate dependent from the smallest applied stresses and since creep is to be observed for the smallest applied stresses, we can conclude that for most geomaterials the yield stress can be considered to be practically zero. The stars at the end of the various curves mark failure. Thus failure is also loading-rate dependent or loading history dependent: with the raising of the loading rate the stress at failure increases, while the strain at failure decreases. Also, in creep tests failure is to be expected at different stress levels (generally much smaller) than in short term tests. All the phenomena described above in conjunction with Fig.1.1 cannot be described by elastic-plastic time-independent constitutive equations. Further we will use rate-type constitutive equations of viscoplastic type to describe such effects.

In Fig.1.1 is also shown that an unloading process exhibits significant irreversible strain components and in most cases hysterezis loops.

1.2 Creep.

All rocks subjected to a constant loading stress-state deform by creep. This was already shown in Fig.1.1. However, rocks have some peculiar properties illustrated in Fig.1.2 obtained for marlclay subjected to uniaxial creep tests. Here is shown the creep as

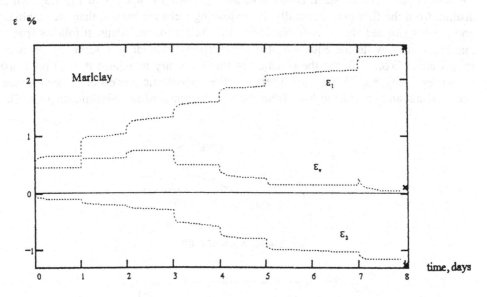

Fig.1.2. Creep curves for marlclay rock in uniaxial tests. For small loading stress
     the volumetric strain exhibits compressibility, while for large stresses -
     dilatancy.

revealed by the axial ε_1 and circumferential ε_2 strains, when stress is increased in successive steps. After each stress increase, the two strains vary by creep (ε_1 increases and ε_2 decreases). The volumetric strain ε_v however, shows compressibility at small stress levels and dilatancy at higher stress levels. This is a typical property of rocks and rock-like materials.

A question may arise: how much is a rock deforming by creep? Or more precise: is the deformation by creep ending at a stable-state, i.e. the creep is always transitory ? Or maybe the creep continues indefinitely under a constant stress: stationary creep if the rate of deformation components are constant, or tertiary creep if the deformation by creep is accelerated ?

In conjunction with these questions we can discuss also the following problem: is the nonlinear constitutive equation to be used with geomaterials, of viscoplastic or viscoelastic nature? Both models can describe creep. To answer to this question one can make uniaxial creep tests with increasing axial stress loading steps (Fig.1.3). If all the increasing steps are $\Delta\sigma_1$, say, we can use the data to determine a stabilization boundary. The procedure is the following: we first apply the stress $\Delta\sigma_1$ and we keep it constant until a stabilization is obtained (transient creep). We now increase again the loading stress up to $2\Delta\sigma_1$, and we determine another point belonging to the stabilization boundary, and so on. By this procedure we can determine the whole stabilization boundary. If we repeat the procedure, using this time, at each additional loading a double stress magnitude $2\Delta\sigma_1$, say, we will obtain another stabilization boundary (shown by squares in Fig.1.3), which is distinct from the first one. Generally, if the loading steps are bigger, then the stabilization boundary is situated above those obtained with smaller loading steps. It follows from here that one cannot write the equation of the stabilization boundary in terms of stresses and strains alone $F(\sigma,\varepsilon) = 0$, but the equation of this boundary must also depend on a history parameter κ, say, e.g. $G(\sigma,\varepsilon,\kappa)=0$. Therefore, the viscoelastic models are unappropriate for geomaterials and one has to look from the very beginning for a viscoplastic one. The

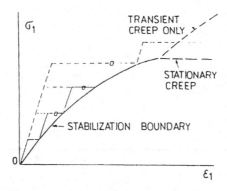

Fig.1.3 History dependence of stabilization boundary and possible description of stationary creep at high stresses.

concept of stabilization boundary (CRISTESCU [1979,1989a]) is fundamental for the formulation of the model. It has to be defined straightforward for the triaxial stress states, the uniaxial models being illustrative only.

In conjunction with the Fig.1.3, if we find for a certain geomaterial that the stabilization boundary is a strictly increasing curve, then the corresponding material exhibits transient creep only. However, if above a certain stress level the stabilization boundary becomes horizontal, stationary creep is also possible. In this case this stabilization boundary becomes a stabilization boundary for the transient creep only. Other possibilities to take into account both transient and stationary creep will be discussed below having in mind those geomaterials for which stationary creep is to be expected at relatively small applied stresses too.

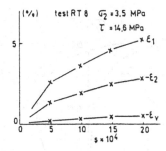

Fig.1.4 Crep curves for Avery Island salt showing volumetric dilatancy (after HANSEN and MELLEGARD [1980]).

From the creep curves for rock salt from Avery Island (see HANSEN and MELLEGARD [1980]) shown in Fig.1.4 obtained in confined tests (the lateral surface of the specimen is subjected to the constant pressure $\sigma_2 = 3.5$ MPa besides the axial stress) it seems that both transient and stationary creep are possible. In this test the volumetric creep produces dilatancy. Thus, generally, the constitutive equation to be formulated has to describe transient, stationary and maybe, tertiary creep as well.

1.3 Dilatancy and/or compressibility

The mechanical properties of geomaterials are distinct from those of other materials mainly concerning the volumetric behavior. Geomaterials always possess pores and/or microcracks which during a loading program may either open or close, may multiply, the pores sometimes collapse to become microcracks, etc. The experimental data mentioned above have shown already some of the features. Let us give some additional details.

If a relatively low porosity rock is subjected to a short-term uniaxial stress test the results obtained are similar to those shown in Fig.1.5 obtained for sandstone. If we follow the curve for the axial strain ε_1 we can distinguish three distinct portions. The nonlinear portion at low stresses corresponds to the closure of microcracks favorably oriented with respect to the applied stress (microcracks belonging to planes nearly orthogonal to the direction of the applied stress). The second, nearly linear portion of the curve has a slope close to the "elastic" one, and no significant changes in the microcrack density occur. In the last nonlinear portion of the curve, other microcracks are opening; their number increases and ultimately failure takes place. The σ_1-ε_v curve also shown in Fig.1.5 exhibits the same three portions, sometimes difficult to pinpoint: in the first one, there where the slope is smaller than the elastic one, the rock is irreversibly compressible, in the second one the volume variation is close to the elastic one and in the third part the rock is dilatant (here the slope of the σ_1-ε_v curve is bigger than the elastic slope or even negative) (see also BRACE et al.[1966]).

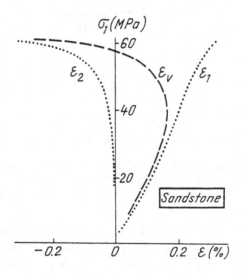

Fig.1.5   Stress-strain curves for sandstone obtained in uniaxial short-term tests, showing volumetric compressibility at low stresses and dilatancy at higher stresses.

If the rock is subjected to various loading rates, the influence on the volumetric behavior is quite sensitive. In Fig.1.6 after SANO et al. [1981], are shown stress-strain curves for granite obtained with four rates of deformation: $4.27 \times 10^{-5} s^{-1}$, $4.14 \times 10^{-6} s^{-1}$, $3.68 \times 10^{-7} s^{-1}$ and $3.42 \times 10^{-8} s^{-1}$. While the change in the strain rate has apparently little effect on the axial strain, the radial and volumetric strain are significantly influenced. With increasing loading rate: the apparent dilatancy threshold raises, the strength increases, the ultimate dilatancy (volumetric strain at failure) decreases. The authors have also shown that

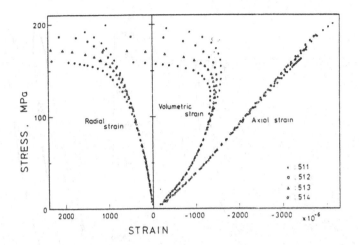

Fig.1.6 Influence of the loading rate on the stress-strain curves for granite
(SANO *et al.* [1981])

there is a quite clear relationship between irreversible volumetric strain and cumulative number of acoustic emissions, and that this relation depends on the strain rate, showing that rock dilatancy is attributable to small-scale cracking events (see also LAJTAI *et al.*[1991]).

The influence of the loading rate on irreversible rock compressibility is less pronounced. However, if the loading rate is changed with several orders of magnitude, in most cases the compressibility is curtailed at very high loading rates.

A specific property of geomaterials is the hydrostatic volumetric compressibility. If a geomaterial is subjected to an increasing hydrostatic pressure, the volume diminishes in an irreversible way (see Fig.1.7). In Fig.1.7 (left) are shown (CONSTANTINESCU [1981]; see also CRISTESCU [1989a]) three uniaxial hydrostatic (confined) compressibility curves for cement concrete, obtained with three loading rates (6.4×10^{-4} s^{-1}, 1.67×10^{-3} s^{-1}, 2.55×10^{-3} s^{-1}). Besides the strong influence of the loading rate, in Fig.1.7 is shown a significant volumetric creep obtained for the constant applied stress $\sigma_1 = 1.166$ GPa during 102 min, 87 min and 70 minutes, respectively. It seems that the first reported hydrostatic volumetric compressibility of a rock is that for schist (Fig.1.7 right) reproduced after CRISTESCU [1979] (see also CRISTESCU and SULICIU [1982]). These two curves have been obtained in uniaxial confined (a lateral displacement is impossible) tests with loading speeds of 4 tons/min and 1 ton/min. The hydrostatic volumetric creep of this hard rock was obtained for $\sigma_1 = 0.372$ GPa and $\sigma_1 = 0.75$ GPa and has been observed a few minutes only. Hydrostatic volumetric creep of rock salt was reported by WAWERSIK and HANNUM [1980].

Temperature has also an influence on both compressibility and dilatancy as it is

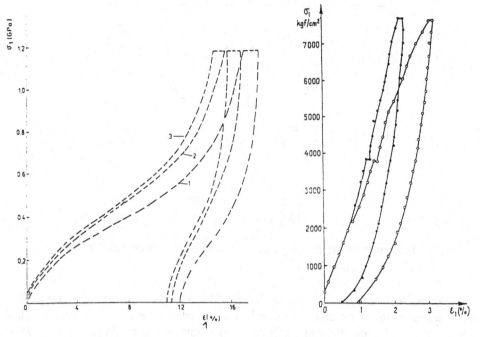

Fig.1.7 Volumetric uniaxial confined loading/unloading curves for cement
        concrete (left) and schist (right), show loading rate influence and
        hydrostatic volumetric creep.

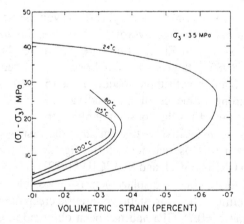

Fig.1.8 Temperature influemce on volumetric compressibility and dilatancy of
        rock salt (after HANSEN and MELLEGARD [1979]).

shown for instance, in Fig.1.8 for Avery Island domal salt in test done by HANSEN and
MELLEGARD [1979] for a confining pressure of 3.45 MPa. The influence of temperature
on creep will be discussed at Chap.6..

For the influence of humidity on deformability of rocks see: LAJTAI *et al.*[1987],
KRANZ *et al.*[1989], CARTER *et al.* [1990], HUNSCHE and SCHULZE [1993],
SCHULZE [1993].

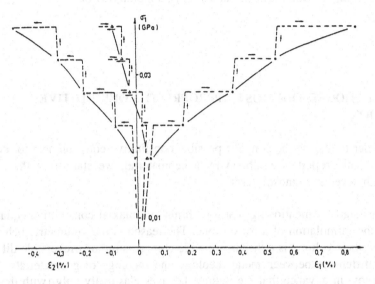

Fig.1.9   Incremental creep curves for sandstone showing oscillatory volumetric
          creep.

Another phenomenon related to dilatancy and compressibility is the oscillatory
behavior of the volume during creep (CRISTESCU [1982]). This is shown for incremental
uniaxial creep tests for sandstone in Fig.1.9. Generally, in the first time interval which
follows the loading the volumetric strain decreases by creep and afterwards increases before
creep stabilization. This oscillation is shown by small arrows in Fig.1.9. Thus both
phenomena, compressibility and dilatancy, seem to be simultaneously present during creep,
but are dominant in successive time-intervals. In other words, it seems that for sandstone,
microcrack formation is sometimes a faster phenomenon than microcrack closure. Below
this phenomenon will be disregarded.

1.4 Historical note.

It is quite difficult to make a history of the development of rheology of geomaterials,
since the concepts and models have steadily and slowly emerged from the experimental
data and from field observations. Some historical references are mentioned in the books by

VUTUKURI *et al.*[1974], LAMA and VUTUKURI [1978], CRISTESCU [1989a], and in several papers or review papers by DRAGON and MROZ [1979], CHANG and YANG [1982], LUX and HEUSERMANN [1983], CARTER and HANSEN [1983], DESAI *et al.*[1986], DESAI and VARADARAJAN [1987], DESAI and ZHANG [1987], GATES [1988a,b], AUBERTIN *et al.*[1991], SENSENY *et al.*[1992], VAN SAMBEEK *et al.*[1992], CARTER *et al.* [1993], CRISTESCU [1993a,b], MUNSON and WAWERSIK [1993], and many others. Some of other related papers will be mentioned below.

2. HOW TO CHOOSE THE MOST APPROPRIATE CONSTITUTIVE EQUATION

In order to choose the simplest possible constitutive equation able to describe the main mechanical properties exhibited by a geomaterial, we start from the information obtained with several diagnostic tests.

Let us start by mentioning that a preliminary uniaxial constitutive equation is not enough for the formulation of a triaxial one. The reason is the volumetric behavior of all geomaterials: their volume is either irreversible compressible or irreversible dilatant. That is the main difference between metal rheology and rheology of geomaterials. With most metals we know in advance that the volume behaves elastically only (with the exception maybe of states close to failure); therefore if we establish a uniaxial constitutive equation (called in plasticity theory the "universal stress-strain curve"), we can rewrite it easily in invariants and thus obtain via this procedure the triaxial constitutive equation. Generally that is no more possible with geomaterials. For instance, if we obtain a uniaxial stress-strain curve for a certain particular geomaterial, the uniaxial strain ε_1 can be expressed in terms of invariants as $\varepsilon_1 = \bar{\varepsilon} + \varepsilon_v/3$, with $\bar{\varepsilon}$ the equivalent strain

$$\bar{\varepsilon} = (\frac{1}{2}\boldsymbol{\varepsilon} \cdot \boldsymbol{\varepsilon})^{\frac{1}{2}} \tag{2.1}$$

and ε_v the volumetric strain. Thus from uniaxial tests we cannot distinguish the amount of contribution of $\bar{\varepsilon}$ and that of ε_v to the total magnitude of ε_1. Moreover, ε_v can increase and decrease during monotonic increase of ε_1. In conclusion, uniaxial tests must be considered preliminary informative tests only. If we make the (strong) assumption that the mechanisms producing the change in shape and those producing volumetric changes are not coupled, then stress-strain curves obtained in true triaxial tests for a single confining pressure, may be useful in order to establish a triaxial constitutive equation.

The approach followed here is to formulate directly a triaxial constitutive equation. For this purpose we will use the information furnished by various diagnostic tests. After getting a general forrmulation for the constitutive equation, we will discuss how the uniaxial information or the information obtained for a single confining pressure can make precise some of the constitutive coefficients.

First we start from the observation that in most geomaterials can propagate both longitudinal and transverse extended body seismic waves. Therefore the instantaneous, elastic rate of deformation component $\dot{\varepsilon}^E$ must be of the form

$$\dot{\varepsilon}^z = \frac{\dot{\sigma}}{2G} + \left\{\frac{1}{3K} - \frac{1}{2G}\right\}\dot{\sigma}\mathbf{1} \qquad (2.2)$$

where the shear and bulk moduli G and K may depend on stress and strain invariants and maybe, on some damage parameter (history of damage evolution). σ is the Cauchy stress, σ is the mean stress and **1** the unit tensor.

Next step is to see if the considered material possesses time-dependent properties as creep and relaxation. As shown above, one has to expect that most geomaterials may possess such properties. Generally both transient and stationary creep are to be described and maybe also the tertiary creep.

In order to describe transient creep which for very long time intervals ultimately leads to a stable state, one usually uses the bracket

$$\langle A \rangle = \frac{1}{2}(A + |A|) = A^+ \qquad (2.3)$$

with the meaning of "positive part of function A". Thus we are able to describe "elastic" unloading starting from a stable state. For instance we can use for the irreversible part of the rate of deformation due to transient creep

$$\dot{\varepsilon}_T^I = k_T\left\langle 1 - \frac{W(t)}{H(\sigma)} \right\rangle \frac{\partial F}{\partial \sigma} \qquad (2.4)$$

where $H(\sigma)$ is the yield function with

$$H(\sigma(t)) = W(t) \qquad (2.5)$$

the equation of the stabilization boundary (where $\dot{\varepsilon}^I=0$, $\dot{\sigma}=0$) with

$$W(T) = \int_0^T \sigma(t)\dot{\varepsilon}_v^I(t)\,dt + \int_0^T \sigma'(t)\cdot\dot{\varepsilon}^{I'}(t)\,dt = W_V(T) + W_D(T) \qquad (2.6)$$

the irreversible stress power per unit volume used as a work-hardening parameter or internal state variable. Further in (2.4) $F(\sigma)$ is a viscoplastic potential which governs the orientation of $\dot{\epsilon}_T^I$. If F coincides with H we say that the constitutive equation is "associated" to a prescribed yield function H. Otherwise the constitutive equation is said to be "nonassociated". k_T is some kind of "viscosity coefficient"; it may depend on stress and strain states and maybe on a damage parameter describing the history of microcracking and/or history of pore collapse to which the rock was subjected. The factor $k_T \, \partial F/\partial \sigma$ itself can be considered to be a variable viscosity coefficient.

The bracket < > is involved in (2.4) in a linear form, easy to handle in future computations. It may, however, be involved in a nonlinear form, as for instance

$$\dot{\epsilon}_T^I = \frac{k_T}{E}[1-\exp{(\lambda\langle H(\sigma)-W(t)\rangle)}]\frac{\partial F}{\partial \sigma} \tag{2.7}$$

or

$$\dot{\epsilon}_T^I = \frac{k_T}{E}\left\langle\frac{H(\sigma)-W(t)}{a}\right\rangle^n\frac{\partial F}{\partial \sigma} \tag{2.8}$$

if necessary (see CRISTESCU and SULICIU [1982] Chap.2). Here $\lambda>0$, $a>0$ and $n>0$ are constants.

In order to describe stationary creep one can either adapt accordingly the function H (see the discussion in conjunction with Fig.1.3) or we can add to (2.4) an additional term, as for instance

$$\dot{\epsilon}_S^I = k_S\frac{\partial S}{\partial \sigma} \tag{2.9}$$

where $S(\sigma)$ is a viscoplastic potential for stationary creep and k_S is a viscosity coefficient for stationary creep, which may depend on strain invariants and on damage, if necessary. Since the bracket < > is absent in (2.9), the creep described by this formula will last as long as stress is applied. In defining the function $S(\sigma)$ we must, however, formulate the restrictive condition that during creep the volumetric compressibility is possible by transient creep only.

Some of the constitutive functions and coefficients mentioned above may be temperature and humidity sensitive as well. Let us mention also that instead of the irreversible stress work per unit volume we could use some other parameters employed to describe irreversible isotropic hardening, as for instance the irreversible equivalent strain

$$\bar{\epsilon}^I(t) = \sqrt{\frac{1}{2}\epsilon^I(t)\,\epsilon^I(t)}. \tag{2.10}$$

or the irreversible equivalent integral of the rate of deformation tensor

$$\overline{\epsilon}^I(T) = \frac{2}{3}\int_0^T \sqrt{\dot{\epsilon}^I(t)\,\dot{\epsilon}^I(t)}\,dt \ . \tag{2.11}$$

However, both these expressions cannot make distinction between irreversibility produced by compressibility and irreversibility produced by dilatancy. These expressions can be used to describe irreversible isotropic hardening for those geomaterials which are either compressible only or dilatant only. That is the reason why we will use from now on the work hardening parameter $W(t)$ only.

In order to formulate the constitutive equation we make several main assumptions:
- We consider only homogeneous and isotropic rocks. Thus the constitutive functions will depend on stress and strain invariants only, besides, maybe, on an isotropic damage parameter. From all possible stress invariants we assume that of greater significance are the mean stress

$$\sigma = \frac{1}{3}(\sigma_1+\sigma_2+\sigma_3) \tag{2.12}$$

and the equivalent stress $\overline{\sigma}$ or the octahedral shear stress τ

$$\overline{\sigma}^2=\sigma_1^2+\sigma_2^2+\sigma_3^2-\sigma_1\sigma_2-\sigma_2\sigma_3-\sigma_3\sigma_1 \tag{2.13}$$

and

$$\tau = \frac{\sqrt{2}}{3}\overline{\sigma} = \sqrt{\frac{2}{3}II_{\sigma'}}$$

with $II_{\sigma'}$ the second invariant of the stress deviator.
- We assume that the material displacements and rotations are small so that

$$\dot{\epsilon} = \dot{\epsilon}^E + \dot{\epsilon}^I \ . \tag{2.14}$$

- The elastic rate of deformation component $\dot{\epsilon}^E$ satisfies (2.2).
- The irreversible rate of deformation component $\dot{\epsilon}^I$ satisfies either (2.4) or (2.9), or is the sum of these two terms, depending on the need to describe either transient creep or stationary creep only or, finally, both kinds of creep occurring simultaneously.
- The initial yield stress of the rock can be assumed to be zero or very close to it.
- The constitutive equation is valid in a certain constitutive domain bounded by a (short term) failure surface, which will be incorporated in the constitutive equation.

Thus the constitutive equation will be written in the form

$$\dot{\epsilon} = \frac{\dot{\sigma}}{2G} + \left(\frac{1}{3K} - \frac{1}{2G}\right)\dot{\sigma}\mathbf{1} + k_T\left(1 - \frac{W(T)}{H(\sigma)}\right)\frac{\partial F}{\partial \sigma} + k_S\frac{\partial S}{\partial \sigma} \quad . \tag{2.15}$$

Let us examine what kind of properties can describe such a constitutive equation. For the time being we disregarded the last term so that the irreversibility will be due solely to the transient creep. From (2.4) we get for the irreversible volumetric rate of deformation

$$\left(\dot{\epsilon}_V^I\right)_T = k_T\left(1 - \frac{W(t)}{H(\sigma)}\right)\frac{\partial F}{\partial \sigma}\cdot\mathbf{1} \quad . \tag{2.16}$$

Let us assume that at an initial time t_o, the initial stress state (the so called "primary" stress) $\sigma^P = \sigma(t_o)$ is an equilibrium stress state, i.e. $H(\sigma(t_o)) = W^P$, with W^P the value of W for the primary stress state (see CRISTESCU [1989a]). A stress variation from $\sigma(t_o)$ to $\sigma(t) \neq \sigma(t_o)$ with $t > t_o$ produced by an excavation, say, will be called <u>loading</u> if

$$H(\sigma(t)) > W(t_o) \tag{2.17}$$

and three cases are possible depending on which of the following inequalities is satisfied by the new stress state:

$$\frac{\partial F}{\partial \sigma}\cdot\mathbf{1} > 0 \qquad or \qquad \frac{\partial F}{\partial \sigma} > 0 \quad compressibility \tag{2.18}$$

$$\frac{\partial F}{\partial \sigma}\cdot\mathbf{1} = 0 \qquad or \qquad \frac{\partial F}{\partial \sigma} = 0 \quad compress/dilatancy \ boundary \tag{2.19}$$

$$\frac{\partial F}{\partial \sigma}\cdot\mathbf{1} < 0 \qquad or \qquad \frac{\partial F}{\partial \sigma} < 0 \quad dilatancy \tag{2.20}$$

We observe that $\partial F/\partial\sigma\cdot\mathbf{1} = \partial F/\partial\sigma$ if we take into account that F depends on stress invariants. Therefore the behaviour of the volume is governed by the orientation of the normal to the surface $F(\sigma)$=const. at the point representing the actual stress state (see Fig.2.1): if the projection of this normal on the σ-axis is pointing towards the positive orientation of this axis, then we have compressibility, otherwise -dilatancy. There where the normal is orthogonal to the σ-axis, there is no irreversible volumetric changes.

If instead of (2.17), the new stress state satisfies

$$H(\sigma(t)) < W(t_o) \tag{2.21}$$

then an <u>unloading</u> takes place and the response of the geomaterial is elastic, according to
(2.2).

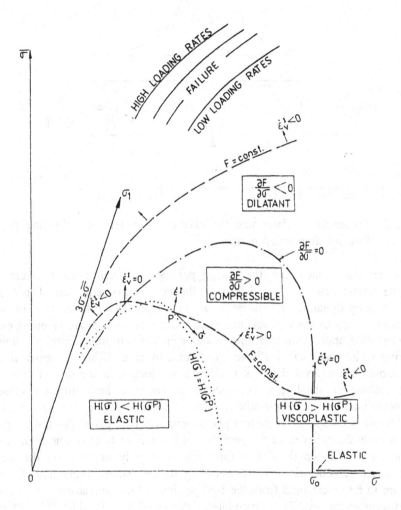

Fig.2.1 Domains of compressibility, dilatancy and elasticity in the constitutive
domain:dash-dot line is the compressibility/dilatancy boundary; failure
depends on the loading rate.

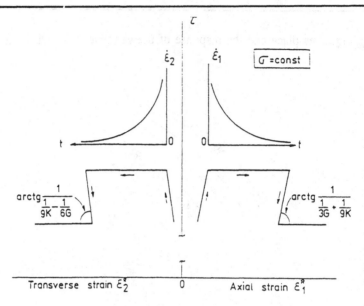

Fig.2.2   Procedure to determine the elastic parameters in unloading processes following short creep periods.

Some remarks concerning the <u>elastic parameters</u>. Their correct determination influences the whole procedure to determine the constitutive equation. These can be determined either by dynamic procedures or by static ones. In the latter case, in order to determine them at various stress levels and since most geomaterials are creeping even for the smallest possible applied stresses, these parameters are determined after a certain time interval during which the rock is allowed to deform by creep. This is shown in Fig.2.2 where at a chosen stress level the rock is allowed to creep until the rate of deformation components become (in absolute value) quite small, ensuring that during the subsequent unloading performed in a comparatively much shorter time interval no significant interference between creep and unloading phenomena will take place. For this purpose, in most cases, a preliminary creep of the geomaterial for half an hour or one hour seems to be quite sufficient. The length of this time interval greatly depends on the kind of geomaterial under consideration. Thus according to the "static" procedure, the elastic parameters are to be determined from the first portions of the unloading slopes obtained after a short-term creep test. This procedure is described by CRISTESCU [1989a]. No complete unloading is to be done since the lower portion of the unloading curves are influenced by other phenomena as well (as the thickness of the specimen, etc.). One can equally well determine the elastic parameters, from the unloading slopes obtained after short-term relaxation tests. It is to be remarked that the values of the elastic parameters determined statically with this procedure compare quite well with those determined dynamically (wave propagation method).

3. TRANSIENT CREEP

Transient creep is described by (2.4). Thus we have to determine from the data the yield function H(σ), the viscoplastic potential F(σ) and the viscosity parameter k_T. We will discuss shortly how these constitutive functions can be obtained.

3.1 The yield function H(σ)

In principle, since the yield function is involved in the equation of the stabilization boundary (2.5), by determining W(t) in creep tests one can determine H(σ). However, since generally triaxial tests are used when dealing with geomaterials, the yield function H(σ) will be determined in two stages corresponding to the two stages of the triaxial test (either "true" or classical, i.e. Kármán). It is natural to look for an yield function of the form

$$H(\sigma) := H_H(\sigma) + H_D(\sigma, \overline{\sigma})$$ (3.1)

where the subscripts come from "hydrostatic" and "deviatoric". Obviously H_D must satisfy the condition $H_D(\sigma,0)=0$.

H_H is determined in hydrostatic tests where $\sigma_1=\sigma_2=\sigma_3$ are increased in a certain chosen time interval $t \in (0,T_H)$. During this stage of the test we determine from the triaxial data

$$W_V(T_H) \equiv W_H(T_H) = \int_0^{T_H} \sigma(t)\,\dot{\epsilon}_V(t)\,dt - \frac{\sigma^2(T_H) - \sigma^2(0)}{2K} \quad .$$ (3.2)

The results are used to determine the stabilization boundary

$$H_H(\sigma) = W_H(t)$$ (3.3)

for hydrostatic loadings. Several examples are given by CRISTESCU [1989a, 1991a,b, 1992, 1993a,b]. In Fig.3.1 is shown an example for rock salt (experimental data by HUNSCHE [1988]). In principle the dependency of H_H on σ must be of the form shown by dotted line in Fig.3.1, i.e. when σ is increasing the existing microcracks are steadily closing so that H_H is a strictly increasing function. When σ increases very much, it approaches a value σ_o, say, which would close all existing microcracks. Thus W_V increases with σ up to a maximum value reached for $\sigma=\sigma_o$, and afterwards W_V remains constant for $\sigma>\sigma_o$ since irreversible volumetric strains are no more possible. For $\sigma>\sigma_o$ the behavior of the volume is elastic. However, when the data available do not cover the whole pressure interval $0 \leq \sigma \leq \sigma_o$, we are sometimes unable to determine with certitude the whole curve shown in Fig.3.1 and therefore the value of σ_o as well. In these cases we determine H_H on a smaller interval of variation of σ, but this function can still be useful in applications.

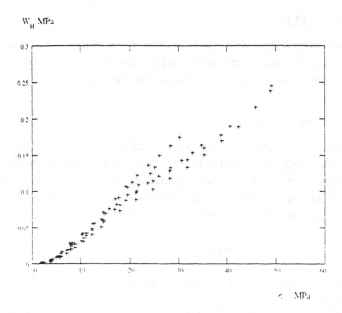

Fig.3.1 Dependence of the hydrostatic stress work on mean stress in hydrostatic
tests.

The data shown in Fig.3.1 can be approximated, for instance, by the empirical
formula

$$H_H(\sigma) := \begin{cases} h_o \sin(\omega \dfrac{\sigma}{\sigma_*} + \varphi) + h_1 & if \quad \sigma \leq \sigma_o \\ h_o + h_1 & if \quad \sigma \geq \sigma_o \end{cases} \qquad (3.4)$$

with h_o=0.116 MPa, h_1=0.103 MPa, ω=2.91°, φ=-64°, σ_o=53 MPa, σ_*=1 MPa for rock salt.
Several other expressions have been used as well for various geomaterials (see
CRISTESCU [1989a, 1991a]).

The function $H_D(\sigma,\tau)$ is to be determined in deviatoric or "nearly" deviatoric tests.
For this purpose the data are used in conjunction with the formula

$$W_D(T) = \int_{T_H}^{T} \boldsymbol{\sigma}'(t) \cdot \dot{\boldsymbol{\varepsilon}}'(t) \, dt - \frac{\boldsymbol{\sigma}'(T) \cdot \boldsymbol{\sigma}'(T)}{4G} + \sigma(T_H) \int_{T_H}^{T} \dot{\varepsilon}_V(t) \, dt \quad (3.5)$$

if the data available are obtained with a true triaxial test apparatus (σ=const in the second
stage of the test), or

$$W_D(T) = \sigma_1(T_H)\,\epsilon_1^R(T) + 2\sigma_2(T_H)\,\epsilon_2^R(T) - \sigma_1(T_H)\left(\frac{1}{3G} + \frac{1}{9K}\right)\sigma_1^R(T)$$

$$-2\sigma_2(T_H)\left(-\frac{1}{6G} + \frac{1}{9K}\right)\sigma_1^R(T) + \int_{T_H}^{T}\sigma_1^R(t)\,\dot{\epsilon}_1^R(t)\,dt - \frac{1}{2}\left(\frac{1}{3G} + \frac{1}{9K}\right)[\sigma_1^R(T)]^2$$

(3.6)

if the data are obtained with a classical (Kármán) kind of apparatus (when $\sigma_2 = \sigma_3 = \text{const.} = \sigma - \tau/\sqrt{2}$). In order to give an example in Fig.3.2 are shown such data obtained for rock salt for various confining pressures shown. These curves can be approximated for instance with expressions of the form

$$H_D(\sigma,\tau) := A(\sigma)\left(\frac{\tau}{\sigma_\star}\right)^{14} + B(\sigma)\left(\frac{\tau}{\sigma_\star}\right)^{3} + C(\sigma)\left(\frac{\tau}{\sigma_\star}\right) \tag{3.7}$$

if the coefficients A, B, and C are thought to depend on the confining pressure σ. Using the data we obtain for these coefficients:

$$A(\sigma) := a_1 + \frac{a_2}{\left(\dfrac{\sigma}{\sigma_\star}\right)^6}\quad,\quad B(\sigma) := b_1\frac{\sigma}{\sigma_\star} + b_2\quad,\quad C(\sigma) := \frac{c_1}{\left(\dfrac{\sigma}{\sigma_\star}\right)^3 + c_3} + c_2 \tag{3.8}$$

with $a_1 = 7\times10^{-21}$ MPa, $a_2 = 6.73\times10^{-12}$ MPa, $b_1 = 1.572\times10^{-6}$ MPa, $b_2 = 1.766\times10^{-5}$ MPa, $c_1 = 26.123$ MPa, $c_2 = -0.00159$ MPa, $c_3 = 3134$.

Thus the yield function is completely determined (not in a unique way). If at this stage of the formulation of the model we accept the idea of an __associated__ constitutive equation (i.e. F≡H) and if only transient creep is of interest, then we have still to determine the viscosity coefficient k_T only, in order to fully determine the constitutive equation. If, for instance, a hydrostatic test is performed, and if the time interval between two successive readings is $t-t_o$, then k_T can be determined using a formula of the form

$$k_T = \frac{\Delta\epsilon_V - \dfrac{\Delta\sigma}{k}}{(t-t_o)\left(1 - \dfrac{W(t_o)}{H(\sigma(t_o))}\right)\Big/\dfrac{\partial H_H}{\partial\sigma}} \tag{3.9}$$

in conjunction with the data. Here Δ is the increment of the corresponding variable during the time interval $t-t_o$ and $H_H(\sigma)=H(\sigma,0)$. However, if any kind of creep tests is available we can use a formula of the form

$$k_t = -\ln\left|1 - \frac{W(t_i)}{H(\sigma)}\right|\Bigg|_{t_i}^{t_f}\frac{H(\sigma)}{\dfrac{\partial F}{\partial\sigma}\cdot\sigma}\frac{1}{t_f - t_i} \tag{3.10}$$

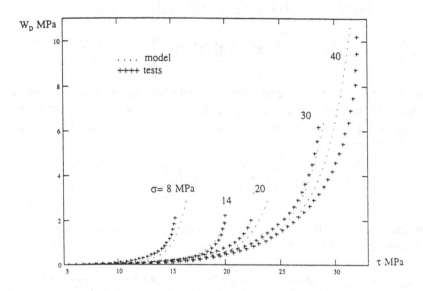

Fig.3.2 Irreversible stress work during deviatoric tests for various confining pressures.

where t_i is the "initial" moment of creep test, t_f is the (arbitrary) "final" moment, $\sigma(t_i)=$const.$=\sigma(t_f)$ is constant during the time interval t_f-t_i, and F is to be replaced by H if the associativeness assumption is accepted. Formula (3.10) has a more general validity since it can be used even if F does not coincide with H.

After determining k_T, the factor E/k_T measured in Poise can be considered to be the "viscosity coefficient". Let us stress here that a correct value of k_T can be obtained in long-term creep tests only. Also k_T may be expected to be variable (possibly depending on damage).

It is quite difficult to decide *a priori* if we have to make the associativeness assumption for a certain geomaterial. That is the reason why we will use a straightforward method to determine the viscoplastic potential. It is only afterwards that we will compare the two functions F and H and also compare with the data the predictions of the two variants of the constitutive equation in order to decide if for that kind of geomaterial, there are circumstances when the associativeness assumption is resonable or not (see also DRUCKER [1988], LADE and KIM [1988], REED and CASSIE [1988], LADE and PRADEL [1990]).

3.2 The viscoplastic potential (CRISTESCU [1991a,1994])

Let us first discuss on which of the stress invariants the function F must depend (as already mentioned, at this stage of the development of the model we disregard the third stress invariant). From the constitutive equation (2.16) we can write

$$\frac{\dot{\epsilon}_1^I}{\dot{\epsilon}_2^I} = \frac{\dfrac{\partial F}{\partial \bar{\sigma}}\dfrac{1}{3} + \dfrac{\partial F}{\partial \sigma}\dfrac{2\sigma_1 - \sigma_2 - \sigma_3}{2\sigma}}{\dfrac{\partial F}{\partial \bar{\sigma}}\dfrac{1}{3} + \dfrac{\partial F}{\partial \sigma}\dfrac{2\sigma_2 - \sigma_3 - \sigma_1}{2\sigma}} \qquad (3.11)$$

if we assume that F depends on $\bar{\sigma}$ and σ alone. If we would further assume $\partial F/\partial \sigma = 0$, then (3.11) becomes

$$\frac{\dot{\epsilon}_1^I}{\dot{\epsilon}_2^I} = \frac{\sigma_1'}{\sigma_2'} \qquad (3.12)$$

i.e. a Saint-Venant type of relation. For uniaxial tests, (3.12) becomes $\dot{\epsilon}_1' = -2\dot{\epsilon}_2'$, i.e. $\dot{\epsilon}_v' = 0$. From here it follows that we cannot have simultaneously $\dot{\epsilon}_1' = 0$ and $\dot{\epsilon}_2' \neq 0$. However, it has been shown by LAJTAI and DUNCAN [1988], in experiments on potash rock that during uniaxial creep tests one can decrease the stress σ_1 up to a value where during creep $\dot{\epsilon}_1' = 0$ but $\dot{\epsilon}_2' \neq 0$ and $\dot{\epsilon}_v' \neq 0$. It seems therefore from the very beginning, that generally the dependence of F on σ cannot be disregarded, mainly if we would like to describe the irreversible volumetric deformation as well.

To determine the viscoplastic potential, we use two fundamental formulae which follow from (2.16) when $H(\sigma) > W(t)$:

$$k_T \frac{\partial F}{\partial \bar{\sigma}} = \frac{\dot{\epsilon}_v^I}{\left\langle 1 - \dfrac{W(t)}{H(\sigma)} \right\rangle} \qquad (3.13)$$

$$k_T \frac{\partial F}{\partial \bar{\sigma}} = \frac{\sqrt{2}}{3}\sqrt{(\dot{\epsilon}_1^I - \dot{\epsilon}_2^I)^2 + (\dot{\epsilon}_2^I - \dot{\epsilon}_3^I)^2 + (\dot{\epsilon}_3^I - \dot{\epsilon}_{31}^I)^2} \left(= k_T \frac{\sqrt{2}}{3}\frac{\partial F}{\partial \tau} \right). \qquad (3.14)$$

In the case of classical triaxial tests the last formula is replaced by

$$k_T \frac{\partial F}{\partial \bar{\sigma}} = \frac{2}{3}\frac{|\dot{\epsilon}_1^I - \dot{\epsilon}_2^I|}{\left\langle 1 - \dfrac{W(t)}{H(\sigma)} \right\rangle}. \qquad (3.15)$$

The determination of $\partial F/\partial \bar{\sigma}$ is done in three stages. First we determine $\partial F/\partial \bar{\sigma}$ for $\bar{\sigma} = 0$ using formula (3.13) and the data. Thus we obtain:

$$k_T \left.\frac{\partial F}{\partial \sigma}\right|_{\bar{\sigma}=0} = k_T \frac{\partial F_H}{\partial \sigma} = \varphi(\sigma) \qquad (3.16)$$

say. Function φ is determined in the stress interval $0 \leq \sigma \leq \sigma_o$ and must possess the properties $\varphi(0)=0$ and $\varphi(\sigma_o)=0$. By definition we take $\varphi(\sigma)=0$ also for $\sigma \geq \sigma_o$, since for $\sigma \geq \sigma_o$ we have $\dot{\varepsilon}_V^I=0$ (no irreversible compressibility is any more possible for $\sigma \geq \sigma_o$ since all microcracks are already closed).

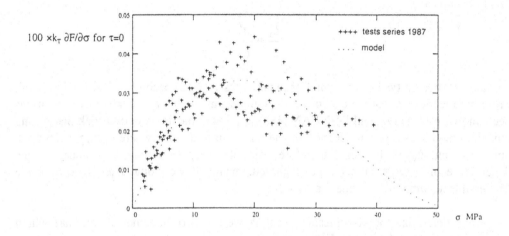

Fig.3.3 Determination of the derivative $\partial F/\partial \sigma$ in hydrostatic tests; dotted line is the prediction of the formulae (3.16) and (3.17) for rock salt.

For instance, the results shown in Fig.3.3 are obtained for rock salt (CRISTESCU and HUNSCHE [1992]). From here we determine

$$\varphi(\sigma) := \begin{cases} q_1 \sigma \, (\sigma - q_2)^2 & if \quad \sigma \leq \sigma_o \\ 0 & if \quad \sigma \geq \sigma_o \end{cases} \qquad (3.17)$$

This function becomes zero for $\sigma = \sigma_o$. In (3.17) we have $q_1 = 1.5 \times 10^{-8}$ s^{-1}, $q_2 = \sigma_o$. Unfortunately, few experimental data were available for high values of σ; thus, the determination of the function $\varphi(\sigma)$ has a certain degree of uncertainty for high pressures. For other geomaterials the function $\varphi(\sigma)$ has been found to have a simpler form, but in some cases it is not determined for the whole interval $0 \leq \sigma \leq \sigma_o$; for instance, for saturated sand (CRISTESCU [1991a])

$$\varphi(\sigma) = h\sqrt{\sigma} \qquad (3.18)$$

with h = 1.07×10^{-3} (kPa)$^{-1/2}$.

In principle, as already mentioned, the function $\varphi(\sigma)$ must be zero for $\sigma \geq \sigma_o$, where σ_o stands for the pressure which closes all microcracks initially existing in the rock. Unfortunately, quite often the available experimental data do not cover the whole interval of pressures up to σ_o. In such cases the function $\varphi(\sigma)$ is determined for a smaller pressure interval (see (3.18)) and the constitutive equation thus determined is valid only for that particular pressure interval; fortunately, such a constitutive equation is still quite often satisfactory for most engineering applications. The example for sand given here is just such a constitutive equation (see below).

Second, we must determine $\partial F/\partial \sigma$ for $\bar{\sigma} \neq 0$. We recall that this function must possess the property of being positive in the compressibility domain and negative in the dilatancy domain. That is why we must first find the geometrical locus in the $\sigma\bar{\sigma}$- plane where the rock passes from compressibility to dilatancy , i.e. the equation of the compressibility/ dilatancy boundary (CRISTESCU [1985d,e]). That is obtained by finding where (i.e. for what stress state) on the σ_1^R- ε_V^R curves obtained in tests performed with various confining pressures σ_2=const., the slope of the tangent is equal to the elastic one ($\dot{\sigma}_1^R = K\dot{\varepsilon}_V^R$, i.e. $\dot{\varepsilon}_V^I$=0; we recall that in the classical triaxial case $\sigma_1^R = \bar{\sigma}$). Here the superscript R means "relative", i.e. stress or strain during the second stage of triaxial test (see CRISTESCU

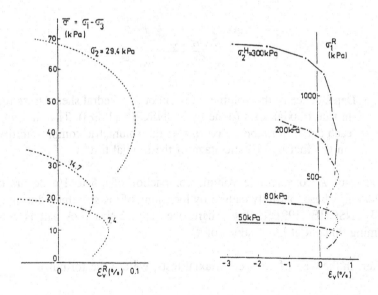

Fig.3.4 Dependence of relative volumetric strain on equivalent stress in triaxial tests for saturated sand (at left) (from the data by LADE *et al.* [1987]). and for dry sand (right) (data from HETLLER *et al.* [1984]).

[1989a]), for which the reference configuration is the state at the end of the first, hydrostatic, stage of the test (when $t=T_H$).

For instance, for saturated sand, using the data obtained with a classical triaxial apparatus by LADE *et al.* [1987] and for dry sand using the data obtained by HETTLER *et al.* [1984], we have plotted the curves shown in Fig.3.4. Volumetric strains are relative, i.e. referred to the beginning of the deviatoric stage of the test. In Fig.3.5 seven similar curves for rock salt are plotted with the data obtained in true triaxial tests by HUNSCHE [1988] (tests in which σ= const. and only τ varies). This time the volumetric strain is total, i.e. the increasing portion of each curve (for τ=0) starts from that point on the ε_V-axis which marks the amount of compressibility obtained at the end of the first hydrostatic stage

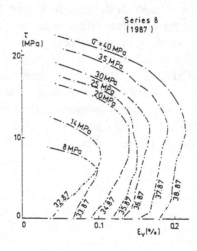

Fig.3.5 Dependence of the volumetric strain on octahedral shear stress for rock salt in true triaxial tests (done by HUNSCHE [1988]). The starting point of each curve from the ε_V-axis marks the volumetric compressibility obtained in the first hydrostatic stage of the triaxial test.

of the triaxial tests. A considerable volume compaction of 0.5 to 1% occurs during the hydrostatic loading phase, partially caused by loosening of the specimen during coring and machining (HUNSCHE [1991a]). From here one can see that rock salt is compressible before becoming dilatant at high values of τ.

Further, in the case of classical triaxial tests, we use the formulae

$$\overline{\sigma} = \sigma_1^R \quad , \quad \sigma = \frac{\overline{\sigma}}{3} + \sigma_2$$

to represent the above locus in the $\sigma\overline{\sigma}$-plane. In the case of "true" triaxial tests this locus is obtained directly from the data (there where the slope of the $\bar{\sigma}$-ε_v^I curves obtained from various tests performed with σ=const. becomes vertical (see Fig.3.5)). Let

$$X(\sigma,\bar{\sigma}) = 0 \qquad\qquad (3.19)$$

be this locus. Sometimes the equation of this locus is written in the form

$$X(\sigma,\bar{\sigma}) := -\left(\frac{\bar{\sigma}}{\sigma_\star}\right)^{\frac{m}{n}} + f\sigma = 0 \qquad\qquad (3.20)$$

with m, n, and f positive constants, and $\sigma_\star$= 1 (the unit stress).

For saturated sand this relation is linear up to σ=50 kPa:

$$X(\sigma,\bar{\sigma}) := -\frac{\bar{\sigma}}{\sigma_\star} + 2f\frac{\sigma}{\sigma_\star} = 0 \qquad\qquad (3.21)$$

with f = 0.562 whereas for rock salt we write the equation of this boundary in the form

$$X(\sigma,\tau) := -\frac{\tau}{\sigma_\star} + f_1\left(\frac{\sigma}{\sigma_\star}\right)^2 + f_2\frac{\sigma}{\sigma_\star} \qquad\qquad (3.22)$$

with f_1= - 0.01697 and f_2=0.8996 (shown as dotted line in Fig,3,6). For sand the form of this boundary for σ>50 MPa is unknown.

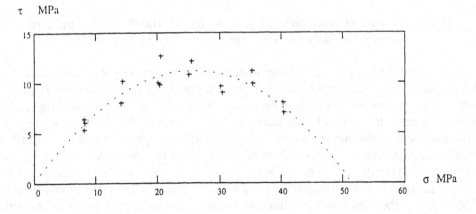

Fig.3.6 Compressibility/dilatancy boundary for rock salt; dotted line is the prediction of (3.22).

In principle, the compressibility/dilatancy boundary must pass through the point $\sigma=\sigma_o$, $\tau=0$, i.e. $X(\sigma_o,0)=0$, and by definition we will accept $X(\sigma,0)=0$ for $\sigma\geq\sigma_o$. If the experimental data available do not cover the whole pressure interval up to $\sigma=\sigma_o$, then we will be unable to determine this boundary up to σ_o, but only one portion of it. Nevertheless, this portion of the boundary may still be useful for the formulation of a model of limited validity (see VAN SAMBEEK [1992] for such a partially determined boundary for rock salt, and formula (3.21) for sand). Generally, the boundary $X(\sigma,\bar{\sigma})=0$ which starts from $\sigma=0$, $\bar{\sigma}=0$ must again reach the σ-axis somewhere for a certain mean stress σ_o and $\bar{\sigma}=0$; otherwise when doing a hydrostatic compression test and the pressures are becoming very high, one will reduce the volume of the rock to a point. For the same reason the so called "cap" models have only a limited validity, for a certain reduced interval of variation of σ (for not too large values of σ).

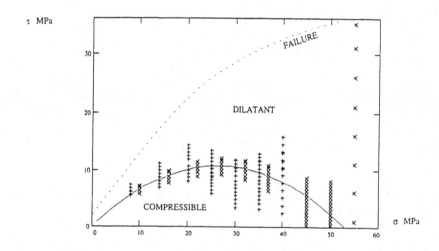

Fig.3.7 Domain of incompressibility in the neighborhood of the compressibility/ dilatancy boundary (++ - experiment; ×× - model prediction).

In conjunction with the compressibility/dilatancy boundary one can make the following remark. If in the neighborhood of the compressibility/dilatancy boundary (along which $\dot{\varepsilon}_v^I=0$) we examine how wide is the domain where ε_v^I has a very small (negligible) variation, we come to the conclusion shown in Fig.3.7 for rock salt. Here by + - lines is marked the width of the domain where the irreversible volumetric strain has a negligible variation (the volumetric strain variation is smaller than the maximum reached on the boundary minus 0.4×10^{-4}). Also by × - lines are shown the same intervals as predicted by the model described below. Therefore we can consider that in a quite wide domain in the neighborhood of the compressibility/dilatancy boundary the volumetric behavior is elastic (incompressible irreversible volumetric strain). This domain of incompressibility enlarges very much at very high pressures. That is the reason why some authors who have disregarded the volumetric compressibility of some geomaterials consider only 2 domains to exist in the constitutive plane $\sigma\bar{\sigma}$: a dilatancy one and an incompressibility one. The last one would be the only domain existing at very high pressures (see for instance CHAN *et al.*[1992]). Also for the same reason it is quite difficult to experimentally determine the exact value of the limit pressure σ_o; it can be determined by extrapolation from data shown in Fig.3.6. Even if σ_o is not accurately determined, it has in principle a great importance for the formulation of the model.

In the next step we have to determine from the data, i.e. from the maxima of the stress-strain curves, obtained in short-term tests, the <u>equation of the failure surface</u>

$$Y(\sigma,\bar{\sigma}) = 0 \qquad\qquad (3.23)$$

assuming that it can be expressed in terms of stress invariants alone. We have found sometimes convenient to write this equation in the form

$$Y(\sigma,\tau) := -r\frac{\tau}{\sigma_*} - s\left(\frac{\tau}{\sigma_*}\right)^6 + \tau_o + \frac{\sigma}{\sigma_*} = 0 \qquad\qquad (3.24)$$

with r, s, τ_o positive constants. For rock salt r=0.91, s=1.025×10⁻⁸, τₒ=1.82, and the surface (3.24) is shown as dotted line in Fig.3.8, together with the data used to determine it (see also HUNSCHE [1992a,b]). Generally, due to the mathematical procedure which follows, it is desirable that the terms containing the mean stress in (3.24) be expressed in the simplest possible form (preferably even in a linear form). As another example, for saturated sand the relation (3.24) is linear (the solid line in Fig.3.10):

$$Y(\sigma,\bar{\sigma}) := -\left(1 + \frac{\alpha}{3}\right)\bar{\sigma} + (2f+\alpha)\,\sigma = 0 \qquad\qquad (3.25)$$

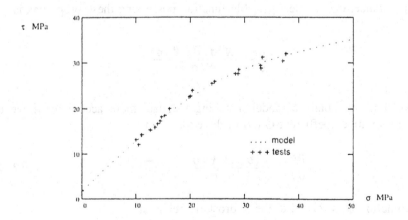

Fig.3.8 Short-term failure boundary for rock salt (dotted line is prediction of (3.24)).

with $\alpha = 1.34$ and f given above, following equation (3.21).

We have now all the ingredients to find an expression for the function $\partial F/\partial\sigma$ for $\tau \neq 0$ which would have all the desired properties, i.e. the change of sign when X is changing sign, and to tend to infinity when Y is tending to zero. That means that $\partial F/\partial\sigma$ would depend on some combination involving the stress invariants (namely X and Y) besides depending maybe individually on σ and $\bar{\sigma}$. In the case of classical triaxial tests we assume that $\partial F/\partial\sigma$ depends on X, Y, σ, $\bar{\sigma}$ and $\sigma_2 = \sigma - \bar{\sigma}/3$ (=const. during the second stage of the test). In the case of true triaxial tests we assume that $\partial F/\partial\sigma$ depends on the same expressions with the exception of σ_2 (this time σ=const. during the second stage of the test). Thus in the case of true triaxial tests we have to determine a function

$$k_T \frac{\partial F}{\partial\sigma} = G(X, Y, \bar{\sigma}, \sigma) \qquad\qquad (3.26)$$

which must possess the following properties:

$$k_T \frac{\partial F}{\partial\sigma}\bigg|_{\bar{\sigma}=0} = G(X, Y, 0, \sigma)\big|_{\bar{\sigma}=0} = \varphi(\sigma) \quad (hydrostatic\ loading)$$

$$
\begin{aligned}
G(X, Y, \bar{\sigma}, \sigma) &> 0 &\leftrightarrow X &> 0 &(compressibility)\\
G(X, Y, \bar{\sigma}, \sigma) &= 0 &\leftrightarrow X &= 0 &(compress/dilat.\ bound.)\\
G(X, Y, \bar{\sigma}, \sigma) &< 0 &\leftrightarrow X &< 0 &(dilatancy)\\
G(X, Y, \bar{\sigma}, \sigma) &\to -\infty &\leftrightarrow Y &\to 0 &(failure)
\end{aligned}
\left.\rule{0pt}{40pt}\right\}\ \bar{\sigma} > 0\ .
\qquad (3.27)
$$

For instance, the simplest possible function possessing these properties is

$$k_T \frac{\partial F}{\partial\sigma} = \frac{X(\sigma, \bar{\sigma})\ \Psi(\sigma)}{Y(\sigma, \bar{\sigma})} \qquad\qquad (3.28)$$

but we may hope to obtain a model matching the data more accurately if we use an additional constitutive coefficient $G_1(\sigma)$ in the expression

$$k_T \frac{\partial F}{\partial\sigma} = \frac{X(\sigma, \bar{\sigma})\ \Psi(\sigma)}{Y(\sigma, \bar{\sigma})}\ [G_1(\bar{\sigma}) + 1] \qquad\qquad (3.29)$$

where the function ψ is related to the hydrostatic behavior

$$\frac{X(\sigma,0)\,\Psi(\sigma)}{Y(\sigma,0)} = \varphi(\sigma) \qquad (3.30)$$

and $G_1(0) = 0$. All functions involved in (3.28) and (3.29) are already determined with the exception of $G_1(\sigma)$ which, in turn, can be determined from experimental data using the formula

$$G_1(\overline{\sigma}) = \frac{\dot{\varepsilon}_v - \dfrac{\dot{\sigma}}{K}}{\left\langle 1 - \dfrac{W(t)}{H(\sigma)} \right\rangle}\, \frac{Y(\sigma,\overline{\sigma})}{X(\sigma,\overline{\sigma})\,\Psi(\sigma)} - 1 \; . \qquad (3.31)$$

For instance for rock salt (3.29) becomes

$$k_T \frac{\partial F}{\partial \sigma} = \frac{\left[-\dfrac{\tau}{\sigma_*} + f_1\left(\dfrac{\sigma}{\sigma_*}\right)^2 + f_2\dfrac{\sigma}{\sigma_*} \right]\left[p_1\left(\dfrac{\sigma}{\sigma_*}\right)^2 + p_2\dfrac{\sigma}{\sigma_*} + p_3 \right]}{-r\dfrac{\tau}{\sigma_*} - s\left(\dfrac{\tau}{\sigma_*}\right)^6 + \tau_0 + \dfrac{\sigma}{\sigma_*}}\, [G_1(\tau) + 1] \qquad (3.32)$$

with $p_1 = -9.83\times10^{-7}$ s^{-1}, $p_2 = -5.226\times10^{-5}$ s^{-1}, $p_3 = 9.84\times10^{-5}$ s^{-1} and

$$G_1(\tau) = u_1\frac{\tau}{\sigma_*} + u_2\left(\frac{\tau}{\sigma_*}\right)^2 + u_3\left(\frac{\tau}{\sigma_*}\right)^3 + u_4\left(\frac{\tau}{\sigma_*}\right)^8 \qquad (3.33)$$

with $u_1 = 0.0365$, $u_2 = -0.00265$, $u_3 = 5.256\times10^{-5}$, $u_4 = 1.57\times10^{-12}$. For rock salt the coefficient $G_1(\tau)$ seems to have sometimes a negligible influence on the accuracy of the prediction obtained with the model. Thus, for the time being we disregard this coefficient, $G_1(\tau)\equiv0$, pending additional data for some other material allowing a careful determination of the function $G_1(\tau)$.

In the third and last stage we must determine $\partial F/\partial\overline{\sigma}$. For this purpose we integrate (3.29) or (3.28) with respect to σ to get

$$k_T F(\sigma,\overline{\sigma}) = \int G(X,Y,\overline{\sigma},\sigma)\, d\sigma + g(\overline{\sigma}) = F_1(\sigma,\overline{\sigma}) + g(\overline{\sigma}) \; . \qquad (3.34)$$

We now differentiate (3.34) with respect to $\overline{\sigma}$, combine the result with formula (3.14) to obtain $g'(\sigma)$:

$$g'(\overline{\sigma}) = \frac{\sqrt{2}}{3} \frac{1}{\left\langle 1 - \frac{W(t)}{H(\sigma)}\right\rangle} \sqrt{(\dot{\epsilon}_1^I - \dot{\epsilon}_2^I)^2 + (\dot{\epsilon}_2^I - \dot{\epsilon}_3^I)^2 + (\dot{\epsilon}_3^I - \dot{\epsilon}_1^I)^2} - \frac{\partial F_1}{\partial \sigma} \quad (3.35)$$

for true triaxial tests, or

$$g'(\overline{\sigma}) = \frac{2}{3} \frac{\dot{\epsilon}_1 - \dot{\epsilon}_2 - \dfrac{\dot{\sigma}_1 - \dot{\sigma}_2}{2G}}{\left\langle 1 - \dfrac{W(t)}{H(\sigma)}\right\rangle} - \frac{\partial F_1}{\partial \sigma} \quad (3.36)$$

in the case of classical triaxial tests (where generally we have $\dot{\sigma}_2 \approx 0$). In these formulae F_1 and H are known functions, while the rate of deformation components are determined from the data.

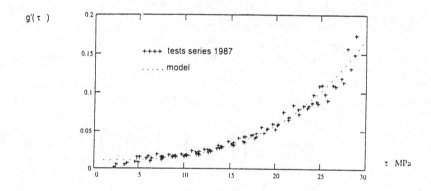

Fig.3.9 Determination of the function g'(τ) from the data obtained for rock salt by HUNSCHE [1991b].

As an example, Fig.3.9 shows the result obtained if the formula (3.35) is used in conjunction with the experimental data obtained by HUNSCHE [1991b] (there are seven tests, each obtained with some other constant value of σ). These data can be approximated by

$$g(\tau) = g_o \frac{\tau}{\sigma_*} + \frac{g_1}{2}\left(\frac{\tau}{\sigma_*}\right)^2 + \frac{g_2}{4}\left(\frac{\tau}{\sigma_*}\right)^4 \quad (3.37)$$

with $g_o = 0.0108$ MPa s^{-1}, $g_1 = 6.582 \times 10^{-5}$MPa s^{-1}, $g_2 = 5.954 \times 10^{-6}$MPa s^{-1} (see Fig.3.9 where the prediction of (3.37) is shown as dotted line). In all these terms, the coefficient g_o plays a significant role: if it is inadequately determined, the function F may exhibit strange properties.

Thus the viscoplastic potential $F(\sigma,\bar{\sigma})$ is entirely determined and therefore the variable viscosity coefficient as well. Sometimes k_T is used as a corrective factor in those cases when the model is used to predict tests or underground deformation lasting a much longer or a much shorter time than in the experimental data which were used to formulate the model. This is done quite easy as will be shown later on. Also k_T may depend on the history of damage evolution.

3.3 Examples.

We will compare the yield functions and the viscoplastic potentials of two geomaterials possessing quite different mechanical properties: saturated sand and rock salt. The constitutive equation for both dry (data by HETTLER et al.[1984]) and saturated sand (data by LADE et al.[1987]) were described by the author elsewhere (CRISTESCU [1991a]). We reproduce here only the expressions for the yield function H and for the viscoplastic potential F for saturated sand, and we show in the $\sigma\bar{\sigma}$-plane the shape of these surfaces. The same will be given for rock salt; the variant of the model is the most recent one after considering additional experimental data and using better programs to analyze them. Early variants have been published by CRISTESCU and HUNSCHE [1992,1993a,b] and CRISTESCU [1993a,b,1994]. A variant for granite was published by CRISTESCU [1987] and presented at the International Symposium "Plasticity Today" held in Udine in 1983 (CRISTESCU [1985e]). From other models given we mention: sandstone (CRISTESCU [1989b]), various kinds of coal (CRISTESCU [1985b], Cristescu et al. [1989]), andesite (CRISTESCU [1985a, 1989a]), bituminous concrete (FLOREA [1993a,b]).

Saturated sand. For saturated sand the function $H(\sigma,\bar{\sigma})$ involved in the equation of the stabilization boundary (let us call it briefly yield function) is of the form

$$H(\sigma,\bar{\sigma}) := a\,\frac{(\bar{\sigma})^7}{\left(\sigma-\dfrac{\bar{\sigma}}{3}\right)^5} + b\bar{\sigma} + c\sigma \qquad (3.38)$$

with $a=4.834\times10^{-7}(kPa)^{-1}$, $b=1.33\times10^{-3}$, and $c=1.0588\times10^{-3}$. The viscoplastic potential has a more involved expression

$$F(\sigma,\bar{\sigma}) := -\frac{2h_1\bar{\sigma}^{-\frac{3}{2}}}{2f+\alpha} + 2fh_1\left(\frac{2}{3}\frac{\sigma^{\frac{3}{2}}}{2f+\alpha} + \frac{2(1+\dfrac{\alpha}{3})\bar{\sigma}^{-\frac{3}{2}}}{(2f+\alpha)^3}\right) +$$

$$\left\{ h_1\overline{\sigma} \frac{\left[(1+\frac{\alpha}{3})\overline{\sigma}\right]^{\frac{1}{2}}}{(2f+\alpha)^{\frac{3}{2}}} - 2fh_1 \frac{\left[(1+\frac{\alpha}{3})\overline{\sigma}\right]^{\frac{3}{2}}}{(2f+\alpha)^{\frac{5}{2}}} \right\} \ln \frac{\left[(2f+\alpha)\sigma\right]^{\frac{1}{2}} + \left[(1+\frac{\alpha}{3})\overline{\sigma}\right]^{\frac{1}{2}}}{\left[(2f+\alpha)\sigma\right]^{\frac{1}{2}} - \left[(1+\frac{\alpha}{3})\overline{\sigma}\right]^{\frac{1}{2}}}$$

$$+ g_o\overline{\sigma} + g_1(\overline{\sigma})^3 \qquad\qquad (3.39)$$

with $h_1 = 2.34\times10^{-3}$ $(kPa)^{-1/2}$, $g_o 0.005$, and $g_1 = 0.62\times10^{-6}(kPa)^{-2}$.

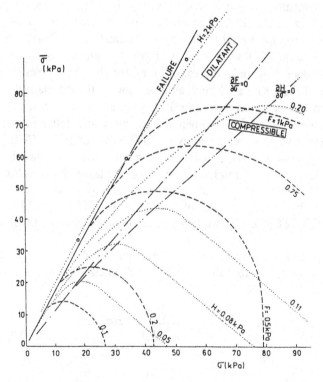

Fig.3.10 Shapes of the surfaces H=const. (dotted lines) and F=const (interrupted lines) for saturated sand; dash-dot line is the compressibility/dilatancy boundary while the failure surface is shown as solid line.

The shapes of several surfaces H=const. (dotted lines) and several surfaces F=const. (interrupted lines) are shown in Fig.3.10. Obviously, these two surfaces are quite distinct. The domain of compressibility is considerably large, while the domain of dilatancy is bounded by the failure surface (solid line) and the compressibility/dilatancy boundary $\partial F/\partial \sigma$ (dash-dot line). It is also obvious that in the whole constitutive domain shown, the associativeness assumption cannot be made. The limit pressure σ_0 was not determined from the existing data, so that the model has a validity limited to the stress states shown in Fig.3.10. It may be that for sand, such large pressures have no practical importance anyway. The meaning of the surface $\partial H/\partial \sigma = 0$ was discussed by CRISTESCU [1991a] in conjunction with the problem of sand stability (sand liquefaction), if the sand is saturated and either undrained or dynamically loaded.

The elastic parameters have not been reported by the authors; from the unloading processes we have determined K=205,300 kPa and afterwards, assuming K=E we have obtained G=77,000 kPa.

Rock Salt. For rock salt the elastic parameters determined from unloading tests are: E=30 GPa, K=21.7 GPa, G=11.8 GPa and υ=0.27. The function H(σ,τ) was already given above (formulae (3.1),(3.4),(3.7), and (3.8)).

The expression for the viscoplastic potential F(σ,τ) is much more involved:

$$k_T F(\sigma,\tau) := \sigma_* \left\{ \frac{f_1 p_1}{4} [Y(\sigma,\tau)]^4 + \left[-\frac{4}{3} f_1 p_1 Z(\tau) + \frac{f_2 p_1 + f_1 p_2}{3} \right] [Y(\sigma,\tau)]^3 + \right.$$

$$\left[3 f_1 p_1 [Z(\tau)]^2 - \frac{3}{2} (f_2 p_1 + f_1 p_2) Z(\tau) + \frac{1}{2} (f_2 p_2 + f_1 p_3 - \frac{\tau}{\sigma_*} p_1) \right] [Y(\sigma,\tau)]^2 +$$

$$\left[-4 f_1 p_1 [Z(\tau)]^3 + 3 (f_2 p_1 + f_1 p_2) [Z(\tau)]^2 - 2 (f_2 p_2 + f_1 p_3 - \frac{\tau}{\sigma_*} p_1) Z(\tau) \right] Y(\sigma,\tau) \cdot$$

$$\left[f_1 p_1 [Z(\tau)]^4 - (f_2 p_1 + f_1 p_2) [Z(\tau)]^3 + (f_2 p_2 + f_1 p_3 - \frac{\tau}{\sigma_*} p_1) [Z(\tau)]^2 - \right. \quad (3.40)$$

$$\left. (f_2 p_3 - \frac{\tau}{\sigma_*} p_2) Z(\tau) - \frac{\tau}{\sigma_*} p_3 \right] \ln Y(\sigma,\tau) + (f_2 p_3 - \frac{\tau}{\sigma_*} p_2) \frac{\sigma}{\sigma_*} \right\} (G_1(\tau)+1) + g(\tau)$$

with Y(σ,τ) defined by (3.24), $G_1(\tau)$ by (3.33), g(τ) by (3.37), and

$$Z(\tau) = -r\frac{\tau}{\sigma_*} - s\left(\frac{\tau}{\sigma_*}\right)^6 + \tau_0 \quad (3.41)$$

Figure 3.11 shows several surfaces H=const (interrupted lines) and several surfaces F=const. (dotted lines); the compressibility/dilatancy boundary is shown as a solid line and the failure surface as a dash-dot line. If we compare this figure with Fig.3.10, we conclude again that the surfaces F=const. and H=const. are distinct but not really so much as in the case of sand. Besides the immediate neighborhood of the failure surface and that of the hydrostatic axis, in the remaining constitutive domain the surfaces F=const. are very close to the Mises kind of surfaces.

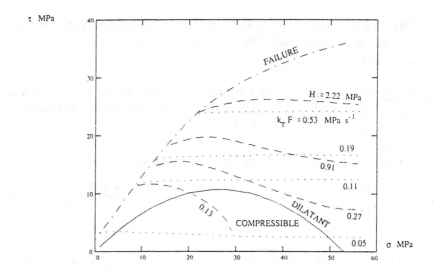

Fig.3.11 Same as for Fig.3.10 but for rock salt.

3.4 Simplified variants of the constitutive equation

From Figures 3.10 and 3.11 it follows that the compressibility/dilatancy boundaries ($\partial F/\partial \sigma = 0$) are considerably distinct from the boundaries $\partial H/\partial \sigma = 0$. Therefore, again, associated constitutive equations are inappropriate for the geomaterials considered, and a significant error will be made in finding the exact position of the compressibility/dilatancy boundary if associated constitutive equations are used. Generally, it seems that for geomaterials an associated constitutive equation can be used only for high pressures, and only seldom elsewhere. However, this has to be judged in conjunction not only with a specific geomaterial, but also with the problem which is to be solved with that particular constitutive equation.

Fig.3.11 also suggests possible simplified variants of the model. If we start from the remark that the surfaces F=const. are quite close to Mises kind of surfaces, we may assume that $\dot{\varepsilon}^I$ is proportional to σ', i.e.

$$F(\boldsymbol{\sigma}) := II_{\sigma'} = \frac{3}{2}\tau^2 \tag{3.42}$$

with

$$\tau^2 = \frac{1}{3}\left(\sigma_1'^2 + \sigma_2'^2 + \sigma_3'^2\right) \quad . \tag{3.43}$$

From here follows

$$\dot{\epsilon}_T^I = k_T\left\langle 1 - \frac{W(t)}{H(\boldsymbol{\sigma})} \right\rangle \boldsymbol{\sigma}' . \tag{3.44}$$

This constitutive equation implies incompressibility, and therefore this simplified model will be unable to describe any irreversible volumetric changes, nor will the short-term failure be incorporated into the constitutive equation. In this case failure has to be considered as an additional condition attached to the constitutive equation. If dilatancy, i.e. microcracking, is of great significance for the problem to be solved, such as in the design of radioactive waste repositories, then the complete model must be used. Recall that in the constitutive equation the function F is involved via the partial derivatives $\partial F/\partial\sigma$ and $\partial F/\partial\tau$ only; the meaning of the terms involved in $\partial F/\partial\sigma$ are obvious from (3.29), while in $\partial F/\partial\tau$ we have two groups of terms: those which were involved in $\partial F/\partial\sigma$ as well and which describe the influence of σ on $\partial F/\partial\tau$, and the function $g'(\tau)$ which is the dominant term in $\partial F/\partial\tau$ (see formulae (3.34) and (3.39),(3.40), where the two groups of terms are already put into evidence). If the simplified constitutive equation (3.44) is used, then k_T is to be determined independently, using formula (3.10) or a similar one.

We may try to simplify the expressions of the function H as well. For this purpose, for rock salt we write this function in the form

$$H(\sigma,\tau) := H_1\tau^{14} + H_2\tau^3 + H_3\tau + H_H \tag{3.45}$$

with the obvious meaning of the coefficients. If for various stress states we estimate the contribution of these four terms, we obtain the results shown in Fig.3.12 for $\sigma=20$ MPa when τ is increased. The position of the compressibility/dilatancy boundary is shown as a dotted line.The relative contribution of the term H_H shown as diamonds, is dominant in the compressibility domain; recall that this term represents the energy of microcracking closing (or opening) when the mean stress varies (either increases or decreases). Further, for higher values of τ some other terms involved in H may become dominant,while close to the failure surface the term H_1 is the dominant one. Thus, the term H_H, completely disregarded by some authors, may be the dominant one, not only during triaxial tests but also in mining applications. That will be shown below in conjunction with the Fig.7.5, Fig.7.6 and Fig.7.7, etc., which show that an underground excavation can produce significant variation of the mean stress in the surrounding rock, and thus significant compressibility or dilatancy are to be expected. Finally, the term H_H plays a more important role in the overall balance of dissipated energies for those geomaterials with a more pronounced initial porosity.

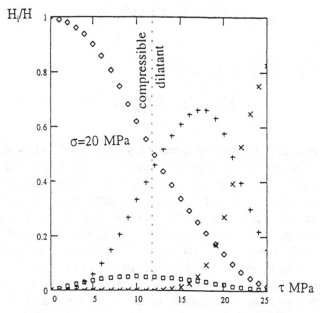

Fig.3.12 Comparison of magnitude of various terms involved in function H during deviatoric tests on rock salt under constant mean stress σ=20 MPa; H_1/H shown by ××; H_2/H by ++; H_3/H by □□; H_H/H by ◊◊.

3.5 Creep formula and comparison with the data

The real test of the model is the comparison of the data with the prediction of the model. First, this can be done, for instance, by trying to reproduce with the model the kind of tests which have been used to formulate the model. We start by establishing the formula describing transient creep. Let us assume that at time t_o the stresses are increased quite fast up to $\sigma(t_o)$ and afterwards are kept constant. We obtain quite easy by integration of the constitutive equation (see CRISTESCU [1989a])

$$1 - \frac{W(t)}{H(\boldsymbol{\sigma})} = \left(1 - \frac{W(t_o)}{H(\boldsymbol{\sigma})}\right) \exp\left[\frac{k_T}{H} \frac{\partial F}{\partial \boldsymbol{\sigma}} \cdot \boldsymbol{\sigma} (t_o - t)\right] \qquad (3.46)$$

where $\sigma(t)=\sigma(t_o)$=const. and $W(t_o)$ is the initial value of W for $t=t_o$. This relation describes the variation in time of W under constant stresses. If the left hand side from (3.46) is introduced in the constitutive equation, by another integration we obtain the <u>creep formula</u>:

$$\epsilon(t) = \epsilon^{\Xi} + \frac{\left(1 - \frac{W(t_o)}{H}\right)\frac{\partial F}{\partial \sigma}}{\frac{1}{H}\frac{\partial F}{\partial \sigma}\cdot\sigma}\left\{1 - \exp\left[\frac{k_T}{H}\frac{\partial F}{\partial \sigma}\cdot\sigma(t_o - t)\right]\right\} \qquad (3.47)$$

with the initial conditions

$$t = t_o : \begin{cases} \epsilon^{\Xi} = 0 \\ \epsilon^{\Xi} = \left(\frac{1}{3K} - \frac{1}{2G}\right)\sigma 1 + \frac{1}{2G}\sigma \end{cases} \qquad (3.48)$$

where $\sigma(t_o)$ is the initial relative stress (reached instantly and taken with respect to the state at time t_o-, when loading is applied -the elastic solution-). Formula (3.47) can be used to describe creep starting from any preliminary existing stress and strain states.

If we would like to describe the stress-strain curves obtained in triaxial tests during the second stage of the test (when σ=const.), then it will be assumed that the stresses are increased by smal successive steps, according to the same law as in the experiments done to establish the model and using the same global loading rate as in the experiments. For each small stress increment the corresponding variation of the strain is obtained from:

$$\Delta\epsilon_1(t) = \Delta\epsilon_1^E(t) + \frac{\left(1 - \frac{W(t_o)}{H(\sigma(t))}\right)\frac{\partial F}{\partial \sigma_1}}{\frac{1}{H}\left(\frac{\partial F}{\partial \sigma}\sigma + \frac{\partial F}{\partial \tau}\tau\right)}\left\{1 - \exp\left[\frac{k_T}{H}\left(\frac{\partial F}{\partial \sigma}\sigma + \frac{\partial F}{\partial \tau}\tau\right)(t_o - t)\right]\right\} \quad (3.49)$$

and similar formulae for the other components. Here all the functions are computed for $\sigma(t)$, and Δ is the variation of the function in the time interval t-t_o. Also both strains and stresses have "relative" meaning, i.e. they are taken with respect to the state existing at the end of the previous loading process. t_o itself is the beginning of a reloading process. Thus these formulae can be used to describe creep tests or any loading histories which can be approximated by successive stepwise stress variations (e.g. tests done with a constant loading rate).

For rock salt (HUNSCHE [1991b]), for the confining pressures σ=35 MPa and 40 MPa, and an approximate loading rate t=21.4 MPa/min, we have obtained the solution shown in Fig.3.13. We have disregarded this time the function $G_1(\tau)$, i.e. $G_1(\tau)$=0, since as shown, the data and the model prediction match quite well even without this function.

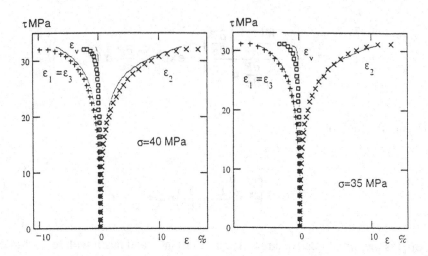

Fig.3.13 Triaxial stress-strain curves for rock salt for σ=const.; model prediction
(solid lines) compared with the data (stars, crosses and squares)

For sand the results are shown in Fig.3.14 (CRISTESCU [1991a]) for both dry sand
(confining pressure in classical triaxial tests σ_2=300 kPa) and saturated sand (σ_2=14.71 kPa).
Here again the matching of the data with the prediction of the model are reasonable.

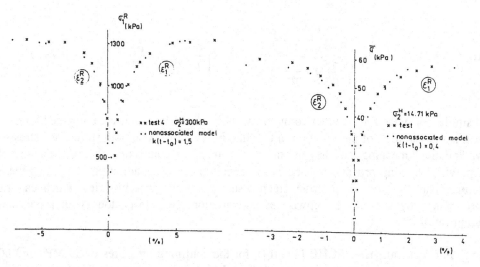

Fig.3.14 Stress-strain curves for dry sand (left) and saturated sand (right); dotted
lines are predictions of the model, and stars are the tests (HETTLER *et
al*.[1984]) for dry sand and (LADE *et al*.[1987]) for saturated sand.

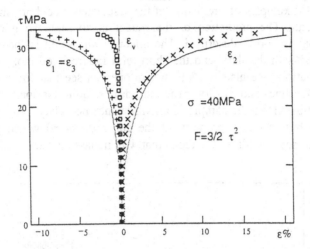

Fig.3.15 Stress-strain curves for rock salt; solid lines are predictions of simplified nonassociated constitutive equation according to (3.42).

Let us consider now <u>simplified solutions.</u> For instance, if we assume for the viscoplastic potential the particular expression (3.42) corresponding to <u>incompressibility</u>, an example of nonassociated model prediction is shown in Fig.3.15 for rock salt. Obviously there is no irreversible volumetric variation though the matching of ε_1- and ε_2-curves are not too bad.

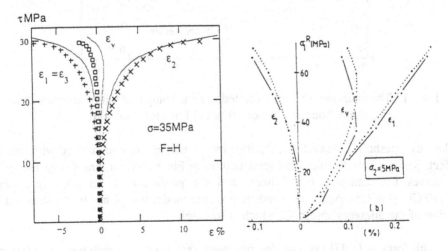

Fig.3.16   Stress-strain curves for rock salt (left - solid lines are predictions of associated constitutive equation - F≡H), and for sandstone (right - CRISTESCU [1989b]).

Another simplified model is the <u>associated</u> one, obtained if we replace the function F by the function H. Examples of prediction of the associated model are shown in the left Fig.3.16 for rock salt. The prediction of the model is no more so good. Mainly the volumetric behaviour is poorly described. The dilatancy threshold is located at too high stresses as compared with the data, and the short term failure is no more incorporated in the constitutive equation. See also Fig.3.17 where one can see that for rock salt the two curves $\partial F/\partial\sigma=0$ (solid line) and $\partial H/\partial\sigma=0$ (dotted line) are quite far apart; that is why one has sometimes the illusion that the compressibility/dilatancy boundary raises and approaches the failure surface, but that is a property of the curve $\partial H/\partial\sigma=0$ which is not the real compressibility/dilatancy boundary. We recall that with nonassociated constitutive equation

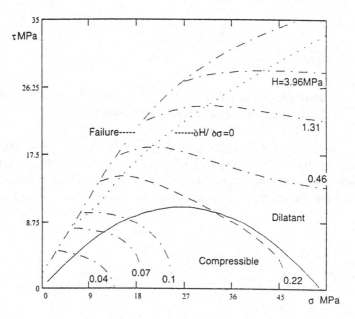

Fig.3.17 The boundary $\partial H/\partial\sigma=0$ (dotted line) as compared with the compressibility/ dilatancy boundary $\partial F/\partial\sigma=0$ (solid line) for rock salt.

the dilatancy threshold is "exactly" described by the model, i.e. coincidence with the data is perfect. Same is true for the short term failure. In Fig.3.16 right, are shown the stress-strain curves for sandstone (solid lines) and the prediction of an associated model (CRISTESCU [1989b]). Again the prediction of the model would not be too bad but the prediction of the dilatancy threshold which is too high.

With formula (3.47) one can describe quite easy <u>transient creep tests</u>. For instance, if for a constant stress $\sigma=40$ MPa, say, we increase in successive steps τ we get the results shown in Fig.3.18. At the first loading step the volumetric creep exhibits a small compressibility followed by dilatancy which is significant for the high loading stresses. To

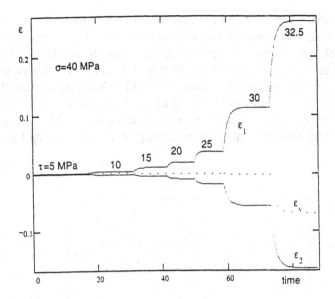

Fig.3.18 Creep curves in true triaxial deviatoric tests.

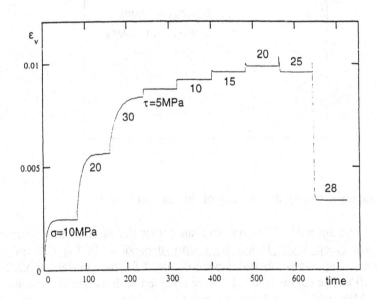

Fig.3.19 Volumetric creep during true triaxial tests showing creep compressibility during hydrostatic test, then elastic behavior and finally dilatancy
creep at high deviatoric stresses.

see in greater details what the prediction of the model would be concerning the volumetric creep, examples have been performed following the loading path of a true triaxial creep test. See Fig.3.19 for such an example. When the mean stress is steadily increased, the volumetric creep exhibits creep compressibility after each loading step. Afterwards, when σ is kept constant and it is τ which is this time increased in successive steps, the model predicts first elastic behavior (with practically no creep) followed by dilatancy by creep which becomes very significant as the stress state approaches failure. Thus, depending on the stress state, the model predicts either compressibility or dilatancy during triaxial creep test.

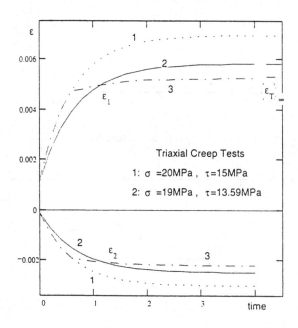

Fig.3.20 Loading history dependence of the transient creep.

Using the same formula (3.47) one can show that the elastic/viscoplastic model describes a transient <u>creep which is loading history dependent.</u> In Fig.3.20 are shown several creep curves. The curves labelled 1 were obtained for the constant applied stress $\sigma=20$ MPa, $\tau=15$ MPa. The curves labelled 2 were obtained with the smaller loading $\sigma=19$ MPa and $\tau=13.59$ MPa, which in a Kármán type of tests means a decrease of axial stress and keeping the confining pressure constant, i.e. $\sigma - \tau/\sqrt{2} = $ const. Finally, the curves labelled 3 were obtained with the first program up to the dimensionless time t=0.6 and afterwards the numerical test continued with the second program (smaller stresses). One can see that the results are distinct.

3.6 Remarks on anisotropy

The development of the elastic/viscoplastic constitutive equations for compressible/dilatant geomaterials as described above is somewhat similar to the historical development of the constitutive equations in classical plasticity theory. Isotropic hardening was considered first, then rate influence. In the present form these models have already been applied to a variety of mining problems. Initial anisotropy or anisotropy induced by irreversible deformation was not yet considered.

The initial anisotropy of rocks and soils, either elastic or inelastic, was considered by BOEHLER [1978], NOVA [1980], SMITH and CHEATHAM [1980], BOEHLER and RACLIN [1982], AMADEI [1983], DESAI [1990], BUDIMAN et al.[1992], and many others. How significant is the initial anisotropy for most rocks is still a debated subject. Some rocks as for instance some coals, are obviously with transverse isotropy.

The anisotropy induced by irreversible deformation of rock salt has been incorporated in a viscoplastic model assuming incompressibility, but considering kinematic hardening and back-stress, by AUBERTIN et al.[1991]. In this formulation the stress σ is replaced by the difference σ-B, where B stands for the back-stress. Also for rock salt, uniaxial models including kinematic and isotropic hardening and again assuming constant volume, were developed by SENSENY et al.[1993].

From a mining engineer point of view, if an underground excavation is done, stresses vary. This variation is generally a loading, and very seldom this loading is followed by a partial unloading. It is only after an infinite time that the new stress state relaxes towards the initial one. A stress-free state is never reached. On the other hand, if closure of caverns or tunnels is of major interest, then the main contribution to closure is that of the stationary creep and not of the transient one (see below §7). That is why from a mining engineer point of view the work-hardening law used is imaterial. Since the isotropic hardening is the simplest possible hardening law and since a stress-free state is never reached in underground works, it seems natural to consider at this stage the isotropic hardening as a satisfactory work-hardening law.

The models developed for induced plastic anisotropy of metals seem not to fit geomaterials, for the same reason for which isotropic work-hardening models for metals are quite distinct from the isotropic work-hardening models for geomaterials (which are strongly dependent on mean stress and which consider irreversible volumetric changes as well). What concerns the induced anisotropy for most rocks, it is primarily due to a peculiar orientation of microcracks towards the direction of maximum applied compressive stress component (see BRACE and BOMBOLAKIS [1963], HORII and NEMAT-NASSER [1985,1986]). As such, the induced anisotropy is strongly dependent on mean stress. Thus a possible kinematic hardening is to be described by a term of the form σ-f(σ)B, where the function f(σ) must possess special properties as for instance, f(σ)→0 when σ increases very much (no anisotropy due to microcracking is possible at very high pressures).

4. STATIONARY CREEP

4.1 The constitutive equation for transient and stationary creep.

It has long been argued that, at least for some geomaterials as for instance rock salt, asphalt, coal and others, one cannot describe their creep with transient creep only. Judging from the significant or even full closure of underground excavations performed in coal or rock salt, which takes place in relatively short time intervals (a few tents of years, say), or from the stress relaxation which occurs sometimes up to very small ultimate stresses, we have to consider for the accurate design of an undeground structure, or for the slow motion of a slope, terms in the constitutive equation which would describe the stationary creep as well. Stationary creep terms may also be important to consider if we would like to describe the slow geophysical motion of the Earth strata.

What concerns the concept of <u>stationary creep</u> we will use the general accepted definition in geomechanics: deformation by creep which takes place without ultimately ending on states belonging to a stabilization boundary, in the sense used in the previous section. Also with stationary creep no equilibrium state is possible unless stress or temperature, say, are significantly varied. Sometimes one caracterizes stationary creep by: under constant (or nearly constant) stress states the <u>irreversible rates of deformation are constant</u>. Such creep processes depends mainly on the stress state (possibly slightly variable) but may depend on temperature, humidity, loading history, etc., as well. During stationary creep the microcrack formation continues. Under unchanged loading conditions, stationary creep continues up to failure or up to the beginning of tertiary creep which ultimately ends in failure.

Further (CRISTESCU [1993a]) we will assume that in order to describe both transient and stationary creep, the irreversible rate of deformation components can be decomposed additively in two components: one describing the transient creep $\dot{\varepsilon}_T^I$ and the other one the stationary creep $\dot{\varepsilon}_S^I$. For the transient creep we will use the formulae given above with the important remark that this time $H(\sigma)=W(t)$ is the equation of the stabilization of the transient creep. Thus

$$\dot{\varepsilon}^I = \dot{\varepsilon}_T^I + \dot{\varepsilon}_S^I = k_T \left\langle 1 - \frac{W(t)}{H(\sigma)} \right\rangle \frac{\partial F}{\partial \sigma} + k_S \frac{\partial S}{\partial \sigma} \qquad (4.1)$$

with S the potential for stationary creep and k_S is the corresponding viscosity coefficient. S is expected to depend on stress invariants while damage may influence assumably the value of k_S.

Let us establish the fundamental formulae necessary for the determination of S from the data and which are equivalent to the formulae (3.13),(3.14). We start from the expression of the rate of deformation components

$$\dot{\epsilon}_1^I = k_T \left\langle 1 - \frac{W(t)}{H(\sigma)} \right\rangle \left(\frac{\partial F}{\partial \sigma} \frac{1}{3} + \frac{\partial F}{\partial \tau} \frac{2\sigma_1 - \sigma_2 - \sigma_3}{9\tau} \right) + k_S \left(\frac{\partial S}{\partial \sigma} \frac{1}{3} + \frac{\partial S}{\partial \tau} \frac{2\sigma_1 - \sigma_2 - \sigma_3}{9\tau} \right) \qquad (4.2)$$

. . . .

for $\dot{\epsilon}_2^I$ and $\dot{\epsilon}_3^I$ the expressions being similar. Here we have assumed that both functions F and S depend on stress via the invariants σ and τ alone. If we take into account the expression of τ

$$\tau = \frac{\sqrt{2}}{3} \left(\sigma_1^2 + \sigma_2^2 + \sigma_3^2 - \sigma_1\sigma_2 - \sigma_2\sigma_3 - \sigma_3\sigma_1 \right)^{\frac{1}{2}} \qquad (4.3)$$

it is easy to get from (4.2) the formulae to be used in conjunction with true triaxial tests:

$$\dot{\epsilon}_V^I = k_T \left\langle 1 - \frac{W}{H} \right\rangle \frac{\partial F}{\partial \sigma} + k_S \frac{\partial S}{\partial \sigma} \qquad (4.4)$$

$$\left[(\dot{\epsilon}_1^I - \dot{\epsilon}_2^I)^2 + (\dot{\epsilon}_2^I - \dot{\epsilon}_3^I)^2 + (\dot{\epsilon}_3^I - \dot{\epsilon}_1^I)^2 \right]^{\frac{1}{2}} = k_T \left\langle 1 - \frac{W}{H} \right\rangle \frac{\partial F}{\partial \tau} + k_S \frac{\partial S}{\partial \tau} \ . \qquad (4.5)$$

In the case of classical triaxial tests the last formula becomes

$$\left| \dot{\epsilon}_1^I - \dot{\epsilon}_2^I \right| = k_T \left\langle 1 - \frac{W}{H} \right\rangle \frac{\partial F}{\partial \tau} + k_S \frac{\partial S}{\partial \tau} \ . \qquad (4.6)$$

Let us consider a typical uniaxial or triaxial creep curve for rock salt, showing the variation in time of one of the strain components under a constant loading (Fig.4.1): the first, nonlinear portion corresponds mainly to transient creep but stationary creep may also be present, while in the last linear one stationary creep is dominant. The open problem is if the transient creep stops at a certain moment or not, i.e it is present in the second part of the creep curve but with a negligible effect.

Before discussing in more detail this subject, let us turn first to a very important aspect which cannot be disregarded (CRISTESCU [1993a]). We use Fig.4.2 for this purpose. The constitutive domain of all possible stress states shown in this figure and bounded by the failure surface can be divided in two subdomains: in one compressibility takes place (X>0) and in the other one dilatancy (X<0), with X=0 the equation of the compressibility/dilatancy boundary (see (3.19)). In the compressibility domain where X>0,

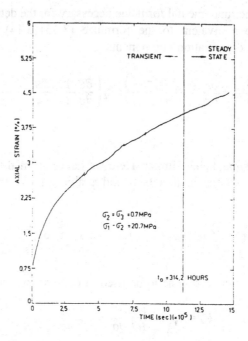

Fig.4.1 Typical creep curve for rock salt (MELLEGARD et al.[1981]) showing a
guessed boundary between transient and stationary creep.

the irreversible volumetric rate of deformation for stationary creep must be zero:

$$\dot{\epsilon}_V^I\big|_S = 0 \qquad\qquad (4.7)$$

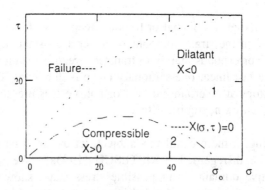

Fig.4.2 Domains of dilatancy and compressibility in a $\sigma\tau$-plane.

since the volume cannot shrink to zero by stationary cree. Thus in the compressibility domain the stationary creep must result in a change in shape only and therefore $\dot\varepsilon_v^I = \dot\varepsilon_v^I|_T$. In the dilatancy domain, where $X<0$, a stationary volumetric creep is possible but producing dilatancy only $\dot\varepsilon_v^I|_S < 0$, besides a change in the shape of the body.

We must now formulate mathematically the above discussed ideas. We have to determine a function $S(\sigma)$ with the following properties:

$$\dot\varepsilon_v^I|_S = k_s \frac{\partial S}{\partial \sigma} := \begin{cases} b\left(\dfrac{\tau}{\sigma_*}\right)^m\left(\dfrac{\sigma}{\sigma_*}\right)^n & if\ X<0 \\ 0 & if\ X\geq0 \end{cases} \qquad (4.8)$$

Here the coefficient $b<0$ may depend on damage, and maybe also on temperature and humidity. The expression $(4.8)_1$ given above was postulated in this form as a generalization of the expressions used by various authors who have formulated uniaxial empirical creep laws based on physical evidence (see CARTER and HANSEN [1983] and WAWERSIK and ZEUCH [1986]). Some other expressions could be considered as well, if necessary. We recall that for rock salt $X=0$ is defined by (3.22) and for saturated sand by (3.21). In a $\sigma\tau$-plane (Fig.4.3) the shape of this surface is shown as a solid line. For a fixed value of $\tau<\tau_m$ we intersect the curve $X=0$ with the horizontal line $\tau=$const. to get

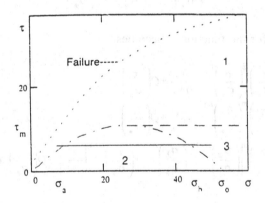

Fig.4.3 Notations and domains used for the determination of the function S.

$$\left.\begin{array}{c}\dfrac{\sigma_a}{\sigma_*}\\[1mm]\dfrac{\sigma_b}{\sigma_*}\end{array}\right\} = \frac{-f_2 \pm \sqrt{f_2^2 + 4f_1\dfrac{\tau}{\sigma_*}}}{2f_1} . \qquad (4.9)$$

We have also $\tau_m = -(f_2^2/4f_1)$. The notations used are shown in Fig.4.3.

If we integrate (4.8) with respect to σ (τ fixed) three cases have to be

distinguished, corresponding to the three regions shown in Fig.4.3. It is easy to obtain

$$
k_s S(\sigma,\tau) := \begin{cases}
b\left(\dfrac{\tau}{\sigma_\star}\right)^m\left(\dfrac{\sigma}{\sigma_\star}\right)^{n+1}\dfrac{\sigma_\star}{n+1}+Q\left(\dfrac{\tau}{\sigma_\star}\right) & if \quad \begin{cases}\tau\geq\tau_m \; or\\ \sigma\leq\sigma_a\end{cases}\\[3mm]
b\left(\dfrac{\tau}{\sigma_\star}\right)^m\left(\dfrac{\sigma_a}{\sigma_\star}\right)^{n+1}\dfrac{\sigma_\star}{n+1}+Q_1\left(\dfrac{\tau}{\sigma_\star}\right) & if \; \sigma_a\leq\sigma\leq\sigma_b \Big\}\tau\leq\tau_m\\[3mm]
b\left(\dfrac{\tau}{\sigma_\star}\right)^m\left[\left(\dfrac{\sigma_a}{\sigma_\star}\right)^{n+1}-\left(\dfrac{\sigma_b}{\sigma_\star}\right)^{n+1}+\left(\dfrac{\sigma}{\sigma_\star}\right)^{n+1}\right]\dfrac{\sigma_\star}{n+1}+Q_2\left(\dfrac{\tau}{\sigma_\star}\right) & if \, \sigma\geq\sigma_b
\end{cases}
$$

$$(4.10)$$

where Q, Q_1, and Q_2 are integration functions in the corresponding domains.

The function S must be continuous at the crossing of the boundary X=0, i.e. for τ=const. at $\sigma=\sigma_a$ and at $\sigma=\sigma_b$ Putting this condition it follows easily

$$
Q_1\left(\frac{\tau}{\sigma_\star}\right)=Q_2\left(\frac{\tau}{\sigma_\star}\right)=Q\left(\frac{\tau}{\sigma_\star}\right) . \qquad (4.11)
$$

Thus the expression for the function S becomes

$$
k_s S(\sigma,\tau) := \begin{cases}
b\left(\dfrac{\tau}{\sigma_\star}\right)^m\left(\dfrac{\sigma}{\sigma_\star}\right)^{n+1}\dfrac{\sigma_\star}{n+1}+Q\left(\dfrac{\tau}{\sigma_\star}\right) & if \quad \begin{cases}\tau\geq\tau_m \; or\\ \sigma\leq\sigma_a\end{cases}\\[3mm]
b\left(\dfrac{\tau}{\sigma_\star}\right)^m\left(\dfrac{\sigma_a}{\sigma_\star}\right)^{n+1}\dfrac{\sigma_\star}{n+1}+Q\left(\dfrac{\tau}{\sigma_\star}\right) & if \; \sigma_a\leq\sigma\leq\sigma_b \Big\}\tau\leq\tau_m\\[3mm]
b\left(\dfrac{\tau}{\sigma_\star}\right)^m\left[\left(\dfrac{\sigma_a}{\sigma_\star}\right)^{n+1}-\left(\dfrac{\sigma_b}{\sigma_\star}\right)^{n+1}+\left(\dfrac{\sigma}{\sigma_\star}\right)^{n+1}\right]\dfrac{\sigma_\star}{n+1}+Q\left(\dfrac{\tau}{\sigma_\star}\right) & if \; \sigma\geq\sigma_b
\end{cases}
$$

$$(4.12)$$

The three expressions given above correspond to the three regions shown in Fig.4.3 as 1, 2 and 3.

We now derivate (4.12) with respect to τ, taking into account (4.9) as well, to get

$$
k_s\frac{\partial S}{\partial\tau} \equiv \frac{bm}{n+1}\left(\frac{\tau}{\sigma_\star}\right)^{m-1}\left(\frac{\sigma}{\sigma_\star}\right)^{n+1}+Q'\left(\frac{\tau}{\sigma_\star}\right)\frac{1}{\sigma_\star} \quad if \; \begin{cases}\tau\geq\tau_m \; or\\ \sigma\leq\sigma_a,\tau\leq\tau_m\end{cases} \quad (4.13a)
$$

$$k_s \frac{\partial S}{\partial \tau} \equiv \frac{bm}{n+1} \left(\frac{\tau}{\sigma_\star}\right)^{m-1} \left(\frac{\sigma_a}{\sigma_\star}\right)^{n+1} - \frac{b}{\sigma_\star} \left(\frac{\tau}{\sigma_\star}\right)^m \left(\frac{\sigma_a}{\sigma_\star}\right)^n \left(f_2^2 + 4 f_1 \frac{\tau}{\sigma_\star}\right)^{-\frac{1}{2}}$$

$$+ Q' \left(\frac{\tau}{\sigma_\star}\right) \frac{1}{\sigma_\star} \qquad if \qquad \sigma_a \le \sigma \le \sigma_b , \ \tau \le \tau_m \tag{4.13b}$$

$$k_s \frac{\partial S}{\partial \tau} \equiv \frac{bm}{n+1} \left(\frac{\tau}{\sigma_\star}\right)^{m-1} \left[\left(\frac{\sigma_a}{\sigma_\star}\right)^{n+1} - \left(\frac{\sigma_b}{\sigma_\star}\right)^{n+1} + \left(\frac{\sigma}{\sigma_\star}\right)^{n+1}\right] - \tag{4.13c}$$

$$\frac{b}{\sigma_\star} \left(\frac{\tau}{\sigma_\star}\right)^n \left(f_2^2 + 4 f_1 \frac{\tau}{\sigma_\star}\right)^{-\frac{1}{2}} \left[\left(\frac{\sigma_a}{\sigma_\star}\right)^n + \left(\frac{\sigma_a}{\sigma_\star}\right)^n\right] + Q' \left(\frac{\tau}{\sigma_\star}\right) \frac{1}{\sigma_\star} \ if \begin{cases} \sigma \ge \sigma_b \\ \tau \le \tau_m \end{cases} .$$

The determination of the constitutive functions and parameters involved in S can be done in the following way. Using formulae (4.4) and (4.8) we can determine b, m, and n from the volumetric creep data available in the dilatancy domain. Afterwards we use formulae (4.5) (or (4.6)) and (4.13) in order to determine the function Q from data in the whole constitutive domain (Any uniaxial or triaxial deviatoric tests with τ varying from zero to failure will do it).

Thus we can obtain the viscoplastic potential function able to describe stationary creep for rocks

$$\dot{\varepsilon}_g^I = k_s \frac{\partial S}{\partial \sigma} = k_s \left[\frac{\partial S}{\partial \sigma} \frac{\partial \sigma}{\partial \sigma} + \frac{\partial S}{\partial \tau} \frac{\partial \tau}{\partial \sigma}\right] \tag{4.14}$$

which are either irreversible dilatant and/or irreversible compressible. This procedure to determine the function S can certainly be applied to any geomaterial, possibly by choosing some other appropriate expressions in (4.8), if necessary.

4.2 Example for rock salt

We will use the data obtained by HERRMANN et al.[1980] for New Mexico rock salt, by HANSEN and MELLEGARD [1980], MELLEGARD et al.[1981], HERRMANN and LAUSON [1981], and SENSENY [1986] for the Avery Island rock salt, in order to determine the constitutive parameters. From these data we get the volumetric behavior shown in Fig.4.4. Using these data in conjunction with formulae (4.4) and (4.8) we obtain b=-1×10^{-14}s^{-1}, m=5 and n=-0.1. Further using formulae (4.6) and (4.13a,b,c) and the data shown in Fig.4.5 we get

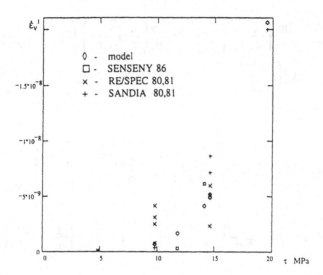

Fig.4.4 Determination of the constitutive constants involved in $\partial S/\partial \sigma$ (see (4.8))
from the data.

$$Q'\left(\frac{\tau}{\sigma_\star}\right) := \frac{p}{\sigma_\star}\left(\frac{\tau}{\sigma_\star}\right)^5 \qquad\qquad (4.15)$$

with $p=3\times10^{-13}$ MPa s^{-1}. The prediction of (4.15) is shown in Fig.4.5 as dotted line. These
estimation are preliminary, pending additional data and a model entirely based on data
obtained for a single kind of rock salt. We remind that the first part of the model (i.e.
functions H and F) are based on data obtained for the Gorleben rock salt, while the second
part (function S) was based on data obtained for the New Mexico and Avery Island rock
salt. For the sake of comparison, we show in Fig.4.6 some results obtained by
MELLEGARD et al.[1981] for Avery Island rock salt in triaxial tests performed with
several confining pressures shown. Using these data and those obtained by HUNSCHE (see
CRISTESCU and HUNSCHE [1992]) we can now compare the compressibility/dilatancy
boundaries and failure surfaces of the Gorleben Rock salt with those of the Avery Island
rock salt. This is done in Fig.4.7. These two kinds of rock salt are certainly not identical;
however, taking into account the possible errors in the data reading and also the fact that
when τ increases the passing from compressibility to dilatancy is very smooth and taking
place in a quite wide interval of variation of τ (as shown in Fig.3.7 it is difficult to
pinpoint this boundary), it seems that the two kinds of rock salt exhibit comparable
mechanical properties. Moreover, the failure boundaries for the two kinds of rock salt are
practically the same (see Fig.4.7). It must be reminded that both compressibility/dilatancy

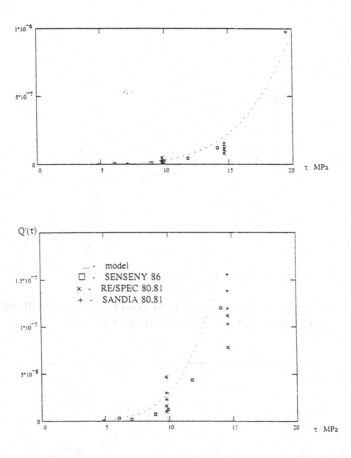

Fig.4.5 Data used for the determination of the function Q and prediction of the
model (dotted line - see (4.15))

boundary and the failure boundary are incorporated in the constitutive equation (in function
F). Certainly, all the other constitutive coefficients for the two kinds of rock salt are also
to be compared. This will not be done any more, since our aim was to give an illustrative
example only, pending a more complete set of experimental data.

With the values of the constitutive parameters given above, function S is fully
determined. In Fig.4.8 are shown several surfaces S=const.Inside the compressibility
domain, the surfaces S=const. do not depend on σ (horizontal straight lines in the $\sigma\tau$-plane).
In the dilatancy domain the surfaces S=const. are slightly conical (i.e. $\partial S/\partial\sigma<0$); here the
surfaces S=const. are dependent only very slightly on σ. This slight dependency on σ is
ensuring, however, that in the compressibility domain no stationary volumetric creep is
possible, while in the dilatancy domain the stationary creep producing dilatancy is possible.

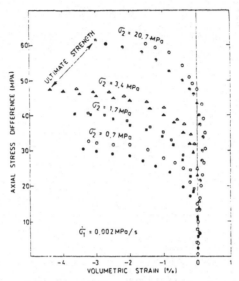

Fig.4.6 Dilatancy and Compressibility of Avery Island rock salt obtained in triaxial tests (after MELLEGARD *et al.*[1981]).

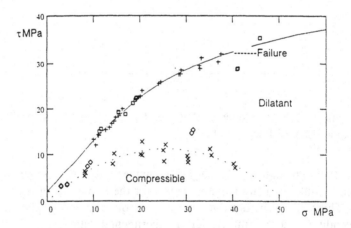

Fig.4.7 Data from Gorleben (×× and ++) and Avery Island (◊◊ and □□) kinds of rock salt, compared for failure surface and compressibility/dilatancy boundary, with prediction of the model (formulae (3.24) and (3.22))

Also in the dilatancy domain the dependency of S on τ is dominant (in both (4.8) and (4.15) τ is involved as a power function, thus $\dot{\varepsilon}_s^I$ can have very big values there where τ has high values; see §7). Moreover, in both domains a stationary creep producing a change in shape is possible.

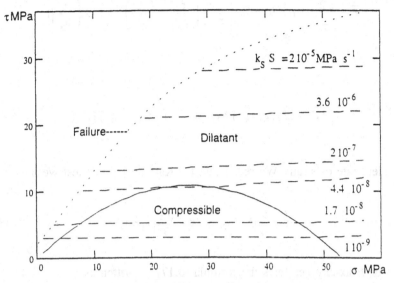

Fig.4.8 Surfaces S=const. shown in a $\sigma\tau$-plane.

4.3 Triaxial generalization of existing stationary uniaxial creep laws.

There are a great deal of uniaxial creep laws well established and which are based either on extensive experimental data or/and physical arguments. We will consider now the following problem: if a certain uniaxial stationary creep formula is already well formulated in the form $\dot{\varepsilon}_1 = f(\sigma_1)$, how can we generalize such formula for triaxial stress state in the general form (4.14), say, and taking into account the volumetric changes as well, if possible. In this paragraph we consider stationary creep only.

The main effort is to determine the function $Q'(\tau)$. Since this function depends on τ alone, it can be determined either from triaxial data or only from uniaxial data. This facilitate very much the determination of $Q'(\tau)$. We first present the formulae necessary to evaluate the data.

The constitutive equation (4.14) for triaxial tests and stationary creep in the dilatancy region (domain 1 in Fig.4.2) is written for one rate of deformation component as

$$\dot{\varepsilon}_1^I = \frac{b}{3}\left(\frac{\tau}{\sigma_*}\right)^m\left(\frac{\sigma}{\sigma_*}\right)^n + \frac{2\sigma_1-\sigma_2-\sigma_3}{9\tau}\left[\frac{bm}{n+1}\left(\frac{\tau}{\sigma_*}\right)^{m-1}\left(\frac{\sigma}{\sigma_*}\right)^{n+1} + Q'\left(\frac{\tau}{\sigma_*}\right)\right] \qquad (4.16)$$

and two similar ones for $\dot{\varepsilon}_2^I$ and $\dot{\varepsilon}_3^I$. If in triaxial tests all rate of deformation components are measured, then from (4.16) we can obtain for Kármán triaxial tests

$$\dot{\varepsilon} = \frac{\sqrt{2}}{3} sign(\sigma_1 - \sigma_2) \left[\frac{bm}{n+1} \left(\frac{\tau}{\sigma_*} \right)^{m-1} \left(\frac{\sigma}{\sigma_*} \right)^{n+1} + Q' \left(\frac{\tau}{\sigma_*} \right) \right] \qquad (4.17)$$

where

$$\dot{\varepsilon}^2 = \frac{2}{9} \left[(\dot{\varepsilon}_1 - \dot{\varepsilon}_2)^2 + (\dot{\varepsilon}_2 - \dot{\varepsilon}_3)^2 + (\dot{\varepsilon}_3 - \dot{\varepsilon}_1)^2 \right] \qquad (4.18)$$

is the equivalent rate of strain. We recall that for Kármán kind of test we have $\varepsilon_2 = \varepsilon_3$ and $\sigma_2 = \sigma_3$, i.e.

$$\dot{\varepsilon} = \frac{2}{3} |\dot{\varepsilon}_1 - \dot{\varepsilon}_2| \qquad \tau = \frac{\sqrt{2}}{3} |\sigma_1 - \sigma_2| \qquad (4.19)$$

For true triaxial creep tests the formula (4.17) is written as

$$\dot{\varepsilon} = \frac{\sqrt{2}}{3} \left[\frac{bm}{n+1} \left(\frac{\tau}{\sigma_*} \right)^{m-1} \left(\frac{\sigma}{\sigma_*} \right)^{n+1} + Q' \left(\frac{\tau}{\sigma_*} \right) \right] \qquad (4.20)$$

if all rate of deformation components are measured. Compared with (4.16) the two formulae (4.17) and (4.20) have the advantage of possessing a single term involving the irreversible volumetric influence and thus being more appropriate if the volumetric data are less accurate.

The function $Q'(\tau/\sigma_*)$ can be determined using either (4.17) if both $\dot{\varepsilon}_1$ and $\dot{\varepsilon}_2$ are measured, or (4.20) if all rate of deformation components are measured, or finally (4.16) if only $\dot{\varepsilon}_1$ is measured. We can handle all cases but the first two are certainly more accurate.

Further we assume that the data can be approximated with the above formulae if various particular expressions are used for $Q'(\tau/\sigma_*)$, as for instance

$$p_0 \left(\frac{\tau}{\sigma_*} \right)^{q_0} \qquad p_1 \left(\frac{\tau}{\sigma_*} \right)^{q_1} + p_2 \left(\frac{\tau}{\sigma_*} \right)^{q_2} \qquad (4.21)$$

$$p \left(\frac{\tau}{\sigma_*} \right)^r sinh \left(q \frac{\tau}{\sigma_*} \right)$$

These expressions are used mainly for rock salt and correspond to the cases when the dislocation climb mechanism is dominant, to the case when an additional unidentified, but experimentally well defined, mechanism is also present, or finally to the case when the slip mechanism is dominant at high stresses. All these mechanisms are thermally activated so that it is expected that some of the constitutive coefficients in (4.20) be temperature and/or humidity dependent.

If only uniaxial data are available for Kármán tests when $\sigma_2 = \sigma_3$ and $\sigma_1 > 0$, and we have also

$$\sigma = \frac{\tau}{\sqrt{2}} \quad , \quad \tau = \frac{\sqrt{2}}{3}\sigma_1 \quad , \quad \sigma = \frac{\sigma_1}{3} \quad ,$$

one can still use formula (4.17) if both rate of deformation components are measured. However, if only $\dot{\varepsilon}_1$ is measured, one has to use (4.16) which for uniaxial tests becomes

$$\dot{\varepsilon}_1 = \frac{b}{3}\left(\frac{\sqrt{2}}{3}\frac{\sigma_1}{\sigma_*}\right)^m \left(\frac{\sigma_1}{3\sigma_*}\right)^n + \frac{\sqrt{2}}{3}(sign\,\sigma_1)\left[\frac{bm}{n+1}\left(\frac{\sqrt{2}}{3}\frac{\sigma_1}{\sigma_*}\right)^{m-1}\left(\frac{\sigma_1}{3\sigma_*}\right)^{n+1} + Q'\left(\frac{\sqrt{2}}{3}\frac{\sigma_1}{\sigma_*}\right)\right]$$

$$(4.22)$$

Using this formula one can determine Q' from uniaxial data. One can use, for instance, the expressions (4.21) for Q', or maybe some other one. However, in these expressions one has to replace τ by $(\sqrt{2}/3)\sigma_1$.

If we would like to disregard the volumetric change, in all the above formulae we have to put b=0. Generally, it is to be expected that the last term Q' from the right hand side of (4.16), (4.17), (4.18) or (4.22) is the dominant one.

Since most of the uniaxial experimentally established creep formulae assume no irreversible volumetric change, it is natural to compare these laws with various variants of (4.22) when the volumetric change was disregarded, or when it is small anyway. Let us give some examples.

First let us consider a power law of the form

$$\dot{\varepsilon}_1 = A\left(\frac{\sigma_1}{\sigma_*}\right)^N \qquad (4.23)$$

for the stationary creep of rock salt. We assume that this law was well formulated from the data so that both A and N are determined. At least A may depend on temperature and humidity. We must compare (4.23) with (4.22) where volumetric change is disregarded (b=0) and we replace Q' with the expression $(4.21)_1$. We obtain easily

$$q_o = N \quad , \quad p_o = \left(\frac{3}{\sqrt{2}}\right)^{q_o+1} (sign\ \sigma_1)\, A \qquad (4.24)$$

for the determination of the coefficients p_o and q_o belonging to the triaxial law, from the ones involved in the uniaxial law A and q. Adding the volumetric changes (quantitatively these are small terms) the triaxial model (4.16) becomes

$$\dot{\epsilon}_1 = \frac{b}{3}\left(\frac{\tau}{\sigma_*}\right)^m\left(\frac{\sigma}{\sigma_*}\right)^n + \frac{2\sigma_1-\sigma_2-\sigma_3}{9\tau}\left[\frac{bm}{n+1}\left(\frac{\tau}{\sigma_*}\right)^{m-1}\left(\frac{\sigma}{\sigma_*}\right)^{n+1} + q_o\left(\frac{\tau}{\sigma_*}\right)^{q_o}\right].\ (4.25)$$

Therefore we have obtained

$$k_s\frac{\partial S}{\partial \sigma} = b\left(\frac{\tau}{\sigma_*}\right)^m\left(\frac{\sigma}{\sigma_*}\right)^n \quad , \quad k_s\frac{\partial S}{\partial \tau} = \frac{bm}{n+1}\left(\frac{\tau}{\sigma_*}\right)^{m-1}\left(\frac{\sigma}{\sigma_*}\right)^{n+1} + p_o\left(\frac{\tau}{\sigma_*}\right)^{q_o}\ (4.26)$$

so that the general constitutive equation of the form (4.14) is entirely determined.

Similarly let us assume that a uniaxial stationary creep law of the form

$$\dot{\epsilon}_1 = B\left(\frac{\sigma_1}{\sigma_*}\right)^2 \sinh\left(C\frac{\sigma_1}{\sigma_*}\right) \qquad (4.27)$$

is well established on the base of uniaxial creep data (thus B and C are well determined). We obtain with the same procedure, this time for the general expression $(4.21)_3$:

$$p = \left(\frac{3}{\sqrt{2}}\right)^3 (sign\,\sigma_1)\, B \quad , \quad q = \frac{3}{\sqrt{2}}\, C \quad , \quad r = 2 \qquad (4.28)$$

which determine the constitutive coefficients p, q, and r , in terms of B and C. This time we get for the general constitutive equation:

$$k_s\frac{\partial S}{\partial \tau} = \frac{bm}{n+1}\left(\frac{\tau}{\sigma_*}\right)^{m-1}\left(\frac{\sigma}{\sigma_*}\right)^{n+1} + p\left(\frac{\tau}{\sigma_*}\right)^2 \sinh\left(q\frac{\tau}{\sigma_*}\right) \qquad (4.29)$$

and again the constitutive equation is fully determined.

A similar procedure can be used for any existing uniaxial stationary creep formula, assuming for Q' an appropriate general expression and comparing with the general formulae shown above. If several creep mechanisms are involved, i.e. several expressions of the form (4.21) (or maybe some other ones) are acting simultaneously, the procedure is the same, the terms being additive if we assume additivity for the rate of deformation components due to several mechanisms.

4.4 The creep formula.

We can follow the procedure described at § 3.5 to get a creep formula for the case when both transient and stationary creep are present. Two variants can be thought.

a. According to the <u>first variant</u> (CRISTESCU [1993a]) both transient and stationary creep are present from the moment of loading but initially the stationary creep is negligible with respect to the transient one. Further both kinds of creep are non-negligible and at a certain moment transient creep ceases. From this moment on the only existing creep process is the stationary one. Let us formulate these ideas in a mathematical form.

From (4.1) by scalar multiplication with σ and integration of the differential equation for W we get

$$1-\frac{W(t)}{H(\sigma)}=-\frac{k_S\frac{\partial S}{\partial \sigma}\cdot\sigma}{k_T\frac{\partial F}{\partial \sigma}\cdot\sigma}+\left[1-\frac{W^P}{H}+\frac{k_S\frac{\partial S}{\partial \sigma}\cdot\sigma}{k_T\frac{\partial F}{\partial \sigma}\cdot\sigma}\right]\exp\left\{\frac{k_T}{H}\frac{\partial F}{\partial \sigma}\cdot\sigma(t_o-t)\right\} \quad .(4.30)$$

Combining this formula with the constitutive equation, by an additional integration we obtain the <u>creep formula</u>:

$$\epsilon(t)=\epsilon^E+\left[-\frac{k_S\frac{\partial S}{\partial \sigma}\cdot\sigma}{k_T\frac{\partial F}{\partial \sigma}\cdot\sigma}\frac{\partial F}{\partial \sigma}+k_S\frac{\partial S}{\partial \sigma}\right](t-t_o)+$$

$$\left[1-\frac{W^P}{H}+\frac{k_S\frac{\partial S}{\partial \sigma}\cdot\sigma}{k_T\frac{\partial F}{\partial \sigma}\cdot\sigma}\right]\frac{H\frac{\partial F}{\partial \sigma}}{\frac{\partial F}{\partial \sigma}\cdot\sigma}\left\{1-\exp\left[\frac{k_T}{H}\frac{\partial F}{\partial \sigma}\cdot\sigma(t_o-t)\right]\right\} \quad (4.31)$$

where ϵ^E is the elastic component. The formula (4.31) can be applied as long as the bracket from (4.1) $<>\neq 0$, i.e. both transient and stationary creep are simultaneously present.

After stabilization of the transient creep, when $<>\approx 0$, only stationary creep is still taking place. Let us denote by t_s the time of stabilization of the transient creep which is obtained from (4.30) for $W(t)\rightarrow H(\sigma)$ as:

$$t_S - t_o = \frac{H}{k_T \frac{\partial F}{\partial \sigma} \cdot \sigma} \ln \frac{k_s \frac{\partial S}{\partial \sigma} \cdot \sigma}{\left(1 - \frac{W^P}{H}\right)\left(k_T \frac{\partial F}{\partial \sigma} \cdot \sigma\right) + k_s \frac{\partial S}{\partial \sigma} \cdot \sigma} \quad . \qquad (4.32)$$

For $t \geq t_S$ the constitutive equation reduces to

$$\dot{\epsilon}^I = k_s \frac{\partial S}{\partial \sigma} \qquad (4.33)$$

and the stationary creep is described by

$$\epsilon(t) = k_s \frac{\partial S}{\partial \sigma}(t - t_S) + \epsilon(t_S) \qquad (4.34)$$

where $\epsilon(t_S)$ is the strain obtained with formula (4.31) at the moment t_S of the stabilization of the transient creep.

In order to give an example let us consider the test RS5504B from HERRMANN et al.[1980], shown by full line in Fig.4.9. With the intention to approximate these data with

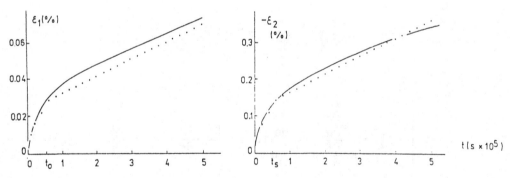

Fig.4.9 Transient and stationary creep curves for rock salt: prediction of the model (dotted line) compared with the data (solid line).

the model, we have to make precise the values of the viscosity coefficients. At this stage of the development of the model, with the data presently available, that can be done only by guessing them by comparing the prediction of the model with the data. For instance if we choose $k_T=3.5\times10^{-3}$ s^{-1} and $k_S=3\times10^{-4}$ s^{-1} we obtain from (4.32) $t_S=58$ days and the prediction of the model (4.31) as shown by dotted line in Fig.4.9. It was not expected to obtain a perfect matching of the data since the model for transient creep corresponds to some other kind of rock salt and also because it is quite difficult to estimate the correct values of the constitutive parameters from small drawings and few tests. The main conclusion is, however, that both transient and stationary creep can be described by the above model.

b. According to the second variant, both transient and stationary creep are always present. In the first time interval after loading transient creep is dominant. It follows a time interval in which neither the transient nor the stationary creep can be disregarded. Finally, in long time intervals transient creep becomes negligible, but no stabilization of the transient creep occurs. Let us formulate these ideas in mathematical terms. We will decompose the irreversible stress work in two terms

$$W(t) = \int_0^T \boldsymbol{\sigma}\cdot\dot{\boldsymbol{\epsilon}}^I dt = \int_0^T \boldsymbol{\sigma}\cdot\dot{\boldsymbol{\epsilon}}_T^I dt + \int_0^T \boldsymbol{\sigma}\cdot\dot{\boldsymbol{\epsilon}}_S^I dt = W_T(t) + W_S(t) \ . \qquad (4.35)$$

We write a formula similar to (3.46) but in which W(t) is replaced by $W_T(t)$ with which we obtain very easy the creep formula:

$$\boldsymbol{\epsilon}(t) = \boldsymbol{\epsilon}^e + \frac{\left(1 - \dfrac{W_T(t_o)}{H}\right)\dfrac{\partial F}{\partial\boldsymbol{\sigma}}}{\dfrac{1}{H}\dfrac{\partial F}{\partial\boldsymbol{\sigma}}\cdot\boldsymbol{\sigma}}\left\{1-\exp\left[\frac{k_T}{H}\frac{\partial F}{\partial\boldsymbol{\sigma}}\cdot\boldsymbol{\sigma}(t_o-t)\right]\right\} + k_S\frac{\partial S}{\partial\boldsymbol{\sigma}}(t-t_o) \ . \quad (4.36)$$

Obviously now the deformation by transient creep stops when $t \to\infty$. However, from (4.36) we can estimate that for $t \ll t_T$ with

$$t_T = \frac{H}{k_T\dfrac{\partial F}{\partial\boldsymbol{\sigma}}\cdot\boldsymbol{\sigma}} \qquad\qquad (4.37)$$

transient creep is dominant. For $t \gg t_T$ the transient creep is negligible. The ultimate value of the transient creep when $t \to\infty$ is

$$\boldsymbol{\epsilon}_{T\infty}^I = \frac{\left(1 - \dfrac{W_T(t_o)}{H}\right)\dfrac{\partial F}{\partial\boldsymbol{\sigma}}}{\dfrac{1}{H}\dfrac{\partial F}{\partial\boldsymbol{\sigma}}\cdot\boldsymbol{\sigma}} \ . \qquad\qquad (4.38)$$

We can use this formula to determine a limit time t_S so that for $t > t_S$ only stationary creep would be non-negligible. This limit time can be obtained from the equality $\varepsilon_S^I(t_S)= \varepsilon_{T\infty}^I$, i.e.

$$t_s = t_o + \frac{\left\langle 1 - \dfrac{W_T(t_o)}{H} \right\rangle \dfrac{\partial F}{\partial \sigma}}{k_s \dfrac{\partial S}{\partial \sigma} \dfrac{1}{H} \dfrac{\partial F}{\partial \sigma} \cdot \sigma} \ . \tag{4.39}$$

From (4.36) follows also the formula for the volumetric creep as

$$\epsilon_V(t) = \frac{\sigma}{K} + \frac{\left\langle 1 - \dfrac{W_T(t_o)}{H} \right\rangle \dfrac{\partial F}{\partial \sigma}}{\dfrac{1}{H} \dfrac{\partial F}{\partial \sigma} \cdot \sigma} \left\{ 1 - \exp\left[\frac{k_T}{H} \frac{\partial F}{\partial \sigma} \cdot \sigma (t_o - t) \right] \right\} + k_s \frac{\partial S}{\partial \sigma} (t - t_o) \ . \tag{4.40}$$

The preliminary comparison of this last variant of the constitutive equation with the data, seems to indicate a better matching. This research is still in progress.

5. DAMAGE AND CREEP FAILURE

As has been shown, for instance, in Fig.1.1, failure of geomaterials is highly loading history dependent. That is why failure conditions formulated in terms of stresses alone would be unable to describe such kind of failures. One has to look for criteria where loading history would also be incorporated.

5.1 Energetic criteria for damage.

In CRISTESCU [1986] a concept of damage for rocks was introduced, which could be used to describe creep failure (see also CRISTESCU [1989a], Chap.9; for models attempting to describe tertiary creep see AUBERTIN et al.[1992], CHAN et al.[1993]). One starts from the idea that the term

$$W_V(T) = \int_0^T \sigma(t)\, \dot{\epsilon}_V^I(t)\, dt \tag{5.1}$$

involved in (2.6), is related to the energy of microcracking, either accumulated by the rock during compressibility period, or released during dilatancy. Thus it is natural to define a scalar damage parameter d related to this concept. A difficulty arises in connection with the correct choosing of the reference configuration for damage, since generally, in any stress-strain state all geomaterials possess some microcracks and/or pores already existing in the corresponding geomaterial. On the other hand, we can imagine tests during which the damage increases during loading and tests during which damage decreases during loading. For instance, in a triaxial deviatoric test (σ=const.) when τ increases, first W_V increases (i.e. microcracks are closing) and then decreases (opening of microcracks). During hydrostatic tests (with τ=0 and σ increases) the damage decreases.

The best reference configuration from where damage could be estimated is the hydrostatic state $\sigma=\sigma_o$, τ=0, corresponding to the smallest pressure which would close all microcracks initially existing in the geomaterial. A typical variation of W_V obtained in a true triaxial test performed for $\sigma<\sigma_o$ is shown as full line in Fig.5.1. First W_V increases during the hydrostatic part of the test when τ=0 and σ increases. Further W_V continues to increases when τ increases and σ=const. The maximum is reached at the crossing of the

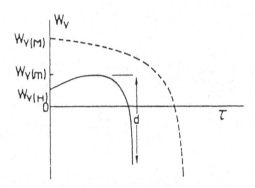

Fig.5.1 Typical variation of W_V during triaxial test for $\sigma<\sigma_o$ (solid line) and for $\sigma>\sigma_o$ (interrupted line).

compressibility/dilatancy boundary and afterwards W_V starts decreasing. Thus the whole decreasing portion of W_V takes place in the dilatancy domain. Moreover, the point of maximum corresponds to the minimum damage and therefore can be considered to be a relative reference configuration for damage. Thus damage can be defined (see Fig.5.1) as

$$d(t) = W_{V(m)} - W_V(t) \qquad (5.2)$$

where the subscript m stands for "relative maximum". A typical variation of W_V in a triaxial test performed for $\sigma>\sigma_o$ is shown in Fig.5.1 as a dotted line. During the hydrostatic part of the test W_V increases and reaches its "absolute" maximum denoted by $W_{V(M)}$. Afterwards W_V is continuously decreasing. According to the above, the damage parameter is a measure of energy release due to microcracking when dilatancy takes place.

The energetic parameter which characterizes the creep failure threshold will be be chosen as the total past energy release due to microcracking during the whole dilatancy process up to failure

$$d_f = W_{v(m)} - W_v(failure) \quad or \quad d_f = W_{v(M)} - W_v(failure) \ . \qquad (5.3)$$

The first term in the right hand side can sometimes be replaced by $W_{v(H)}$ since the difference between $W_{v(H)}$ and $W_{v(m)}$ is in some cases small and, on the other hand, in applications $W_{v(H)}$ can be estimated much easier than $W_{v(m)}$.

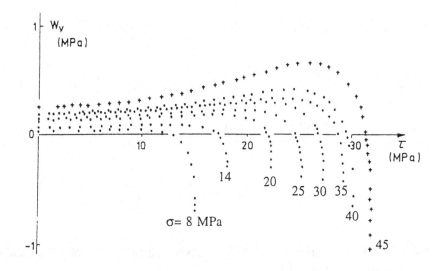

Fig.5.2   Variation of W_v during true triaxial tests performed with several
           confining mean stresses on rock salt.

For instances, in Fig.5.2 is shown for rock salt the variation of W_v in triaxial tests performed with several confining values of σ shown. The value of $W_{v(H)}$ (for $\tau=0$) corresponds to the maximum value of W_v reached during the hydrostatic stage of the test, while the maximum reached when τ is increased corresponds to the dilatancy threshold. If we use now the definition (5.3), we can determine d_f at failure. The results obtained in a great number of tests (due to HUNSCHE [1991b]) is shown in Fig.5.3. Though a significant scatter of data is obvious, it seems that d_f is practically constant

$$d_f = 0.689 \, MPa \ . \qquad (5.4)$$

Also for rock salt HUNSCHE [1991a] reports a value situated between 0.5 and 1 MPa.

For granite the value obtained by CRISTESCU [1986] by analyzing the data of SANO *et al.*[1981] is $d_f = 1.67 \times 10^{-4}$ GPa while for coal $d_f = 4 \times 10^{-3}$ MPa (CRISTESCU 1987). Finally we mention that a more precise estimation of the value of d_f needs quite accurate data and recording of the data up to failure. The recordings must be done in time.

5.2 Creep failure.

We would like to show how the above concept can be used to describe creep failure of geomaterials.

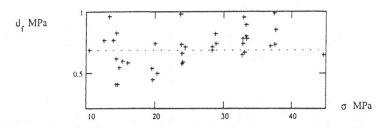

Fig.5.3 Values of ultimate damage parameter for rock salt obtained in a great number of triaxial tests performed with various confining pressures.

Let us consider first failure which takes place in creep tests. We will consider rather high octahedric values for τ, there where creep failure is to be expected after not too excessively long time interval. If τ is only moderately big (close to the compressibility/dilatancy boundary, say), then creep failure is practically no more possible (it may take place after an excessively long time interval). If τ approaches the compressibility/dilatancy boundary, then the time to creep failure tends to infinity. Since in creep tests the applied stresses are kept constant, it is very easy to estimate the integral (5.1). For instance, if we consider transient and stationary creep described by the constitutive law (4.1) and the variant of this model illustrated by the creep formula (4.36), then the volumetric creep is described by (4.40), and (5.1) can be integrated to give

$$W_V(t) = \sigma \epsilon_V^I(t) = \frac{\left\langle 1 - \frac{W_T(t_o)}{H} \right\rangle \frac{\partial F}{\partial \sigma} \sigma}{\frac{1}{H} \frac{\partial F}{\partial \sigma} \cdot \sigma} \left\{ 1 - \exp\left[\frac{k_T}{H} \frac{\partial F}{\partial \sigma} \cdot \sigma (t_o - t) \right] \right\} +$$

$$k_s \frac{\partial S}{\partial \sigma} \sigma (t - t_o) + W_V^P \qquad (5.5)$$

where W_V^P stands for the initial "primary" value of W_V at time t_o. For instance, if we would like to describe a creep test performed during the deviatoric part of a triaxial test and we start the test from the end of a hydrostatic compression creep test which ended at time t_H, say, then $W_v^P=H(\sigma^P,0)$ with σ^P the "primary" value of σ. If we disregard the difference between $W_{V(H)}$ and $W_{V(m)}$ then the damage parameter (5.2) becomes

$$d(t) = W_v^P - W_v(t) \qquad\qquad (5.6)$$

with the right hand term obtained from (5.5). If now d(t) is replaced by its ultimate value d_f, then from (5.6) can be estimated the time when a creep failure is to be expected.

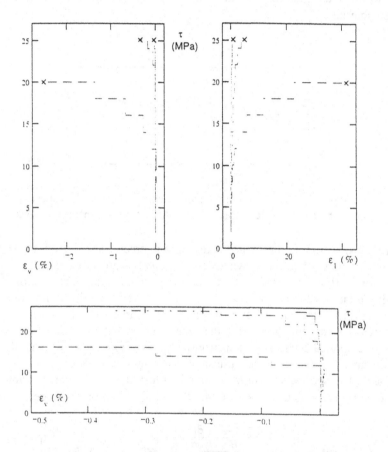

Fig.5.4 Stress-strain curves obtained with three distinct global loading rate, showing
       that an increase of the loading rate is increasing the stress at failure
       and decreasing the strain at failure. On the left and bottom figures:
       volumetric dilatancy is curtailed with increasing loading rate.

As an example in Fig.5.4 are shown several stress-strain curves obtained in true triaxial tests for rock salt with $\sigma_2= 25$ MPa, by increasing the stress in steps of 2 MPa each, and after each such increase the stress is kept constant always for the same interval of time during a single test. In the other tests shown the stress steps are the same, but the time during which the stress is kept constant, is increased from 0.1 to 50 and finally to 5×10^5 (dimensionless time units). From the upper right part of the Fig.5.4 one can see that the general trend of the stress-strain curves is to raise as the global loading rate increases. The stars at the end of various curves mark the failure threshold according to (3.24) for the two upper curves and according to (5.3) and (5.5) for the lower curve. Thus, as the loading rate increases, the stress at failure increases, but the axial strain at failure decreases. The lower part of the Fig.5.4 is showing enlarged (from the upper left figure) how the volumetric strain is varying during the same kind of tests. Thus by increasing the global loading rate, the dilatancy and compressibility are curtailed, i.e. ε_v at failure is smaller in absolute magnitude as the loading rate increases and so is the maximum ε_v .

Similar results have been obtained for other rocks as well. An example for andesite is shown in Fig.5.5 from CRISTESCU [1989a], where the model used was an associated one. The average loading rates are $\dot\sigma$= 6.06×10^{-8} GPa s^{-1}, 6.17×10^{-7} GPa s^{-1}, and 5.7×10^{-6} GPa s^{-1} respectively.

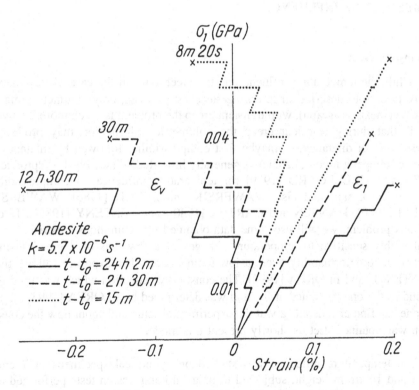

Fig.5.5 Incremental creep tests for andesite: after each stress increase the stress is kept constant for the time interval shown. Stars mark failure, while the corresponding shown time is the time elapsed after the last loading.

The criteria for threshold of creep failure were also used to describe long-term failure around underground excavation (see § 7).

5.3 Tertiary creep.

The tertiary creep is considered to be the result of an advanced stage of damage reached by the geomaterial subjected to a stress state during a long time interval. Due to the damage, most constitutive coefficients may change more or less. It is thought that one of the constitutive coefficients which is highly damage sensitive is the viscosity coefficient k_s. At the beginning of the stationary creep, k_s is considered to be constant, but it becomes strongly dependent on damage as the creep deformation proceeds and approaches tertiary creep and ultimately failure. Thus it is thought that tertiary creep could be described by finding how k_s depends on damage. This subject matter is now under consideration.

6. TEMPERATURE INFLUENCE

6.1 Transient creep.

While the temperature influence on the creep laws in the case of stationary creep is relatively easy to describe since during steady state creep, only volumetric dilatancy is possible (see next paragraph), with transient creep the problem is much more involved. The reason is that during transient creep the volumetric deformation may produce either compressibility or dilatancy, or maybe first compressibility followed by dilatancy. The influence of temperature on dilatancy is generally little studied (see Fig.1.7 reproduced after HANSEN and MELLEGARD [1979] for temperature influence on compressibility and dilatancy of rock salt, and also WAWERSIK and ZEUCH [1986], WAWERSIK and HANNUM [1980], HANSEN and CARTER [1980], and SENSENY [1985]). In order to discuss this problem, we will use some data obtained for bituminous concrete, a rock-like material highly sensitive to temperature changes of a few tens of degree Celsius, with significant creep deformability even at low temperatures. The exposure follows the paper by CRISTESCU and FLOREA [1992]. The constitutive equation which was used is of the same kind as the one presented at § 3 and was determined by FLOREA [1993a,b]. In these papers one can find also details about the experimental setup and about how the constitutive equation was obtained. Let us shortly present this model.

The temperature influence was studied on cylindrical specimens of 7 cm initial diameter and 14 cm in height, subjected to uniaxial compression tests performed at six temperature levels: +50°C, +35°C, +20°C, +8°C, +1°C and -15°C. The testing speed was 0.01 MPa/s. For such testing speeds the temperature of the specimen remains practically constant during the test. Both longitudinal and transverse strains were recorded.

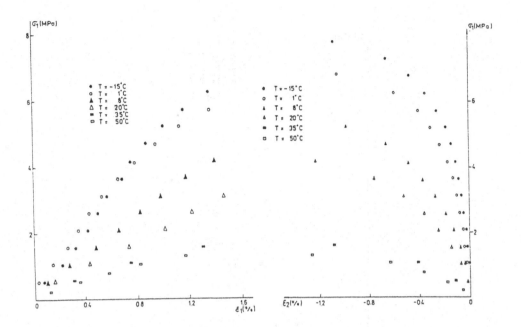

Fig.6.1 Temperature influence on σ_1-ε_1 uniaxial stress-strain curves for bituminous concrete.

Fig.6.2 Temperature influence on σ_1-ε_2 uniaxial curves for bituminous concrete.

The results of the tests are shown in Fig.6.1-6.3. From these figures it is obvious that temperature has a significant influence on the stress-strain curves; the compressive stength varies between 1.5 MPa at 50°C and 8.5 MPa at the temperature -15°C. The volume of the bituminous concrete is strongly compressible at lower stresses but becomes dilatant at higher stresses; both properties are highly temperature dependent.

The temperature influence on the mechanical properties of bituminous concrete was studied also in short-term uniaxial compressive creep tests (where the stress was kept constant at a fixed level during a quarter of an hour). The tests have been carried out up to failure, at the following temperatures: +50° C, +35° C, +18° C, +8° C, 0° C and -12° C. Again both longitudinal and transverse strains were recorded during tests.

In order to determine the constitutive equation, first to be determined is the Young's modulus. From triaxial unloading tests is determined E=4070 MPa, while the mean value of Poisson's ratio was found to be υ=0.288.

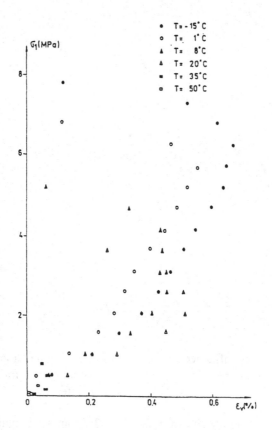

Fig.6.3 Temperature influence on compressibility and dilatancy of bituminous
         concrete in uniaxial tests.

The yield function H was determined as described at § 3, using classical triaxial data:

$$H(\sigma,\overline{\sigma}) := b\left(\frac{\sigma}{\sigma_*}\right)^{\frac{3}{2}} - c\left(\frac{\sigma}{\sigma_*}\right)^{2} + \frac{d}{e+\dfrac{1}{\sigma_*}\left(\sigma-\dfrac{\overline{\sigma}}{3}\right)}\left(\frac{\overline{\sigma}}{\sigma_*}\right)^{4} + \frac{h}{g+\dfrac{1}{\sigma_*}\left(\sigma-\dfrac{\overline{\sigma}}{3}\right)}\frac{\overline{\sigma}}{\sigma_*} \quad (6.1)$$

where b=7.5×10^{-5} MPa, d=0.336×10^{-7} MPa, h=0.035 MPa, c=1.7×10^{-6} MPa, e=30, g=72.99
and σ_*=1 MPa.

Since the matching of the data with the prediction of the associated constitutive equation was not very good, one has arrived at the conclusion that for this material too, a nonassociated constitutive equation is necessary. The viscoplastic potential function for transient creep was determined in the form

$$F(\sigma,\overline{\sigma}) := \Phi(\sigma,\overline{\sigma}) + g(\overline{\sigma}) \tag{6.2}$$

with

$$\Phi(\sigma,\overline{\sigma}) = K_1(\overline{\sigma}) \left\{ -\frac{2\left(\dfrac{\sigma}{\sigma_\star}\right)^{\frac{1}{2}}}{K_o} + \frac{(K_V(\overline{\sigma}))^{\frac{1}{2}}}{K_o^{\frac{3}{2}}} A \right\}$$

$$+ K_2 \left\{ \frac{6 K_V(\overline{\sigma})\left(\dfrac{\sigma}{\sigma_\star}\right)^{\frac{1}{2}} + 2 K_o\left(\dfrac{\sigma}{\sigma_\star}\right)^{\frac{3}{2}}}{3 K_o^2} - \frac{(K_V(\overline{\sigma}))^{\frac{3}{2}}}{K_o^{\frac{5}{2}}} A \right\} \tag{6.3}$$

where

$$K_V(\overline{\sigma}) = \left(\frac{\overline{\sigma}}{\sigma_\star}\right)^{\frac{p}{q}}, \qquad K_2 = F_c \varphi_o, \qquad K_o = F_o + \alpha \tag{6.4a}$$

$$A = \ln\left|\frac{(K_V(\overline{\sigma}))^{\frac{1}{2}} + \left(\dfrac{\sigma}{\sigma_\star} K_o\right)^{\frac{1}{2}}}{(K_V(\overline{\sigma}))^{\frac{1}{2}} - \left(\dfrac{\sigma}{\sigma_\star} K_o\right)^{\frac{1}{2}}}\right|, \qquad K_1(\overline{\sigma}) = \varphi_o\left(\frac{\overline{\sigma}}{\sigma_\star}\right)^{\frac{p}{q}} \tag{6.4b}$$

$$g(\overline{\sigma}) = g_o\frac{\overline{\sigma}}{\sigma_\star} + g_1\left(\frac{\overline{\sigma}}{\sigma_\star}\right)^3 \tag{6.4c}$$

with $\varphi_o = 9.55 \times 10^{-5}$ MPa, $p/q = 1.7$, $F_c = 9.77$, $\alpha = 52.33$, $g_o = 3.6 \times 10^{-4}$ MPa, and $g_1 = 10^{-7}$ MPa.

Checking qualitatively the prediction of the model as compared with the data one has found that from all constitutive parameters involved in the elastic/viscoplastic nonassociated constitutive equation, those which are the most temperature sensitive are the viscosity coefficient k and the coefficient F_c related to the dilatancy threshold.

In order to determine the function describing the dependence of the coefficient F_c on temperature, we use the equation of the compressibility/dilatancy boundary (FLOREA [1993a,b]):

$$\left(\frac{\overline{\sigma}}{\sigma_*}\right)^{\frac{p}{q}} = F_c \frac{\overline{\sigma}}{\sigma_*} \quad . \tag{6.5}$$

We particularize this relation for the case of unconfined uniaxial compressive tests to get:

$$F_c = 3\left(\frac{\sigma_1}{\sigma_*}\right)^{0.7} \quad . \tag{6.6}$$

Introducing in this formula the value of the stress σ_1 corresponding to the dilatancy threshold obtained in uniaxial unconfined compression tests, performed at various temperatures, we obtain a linear dependency of the coefficient F_c (a dimensionless quantity) on temperature T (in Celsius degrees)

$$F_c(T) = F_1 T + F_o \tag{6.7}$$

with F_1= -0.835 (°C)$^{-1}$ and F_o=43.903.

Thus, the expressions of the constitutive coefficients K_o and K_2 (see (6.4)), which depend on F_c, become functions of temperature:

$$\begin{aligned} K_2(T) : &= F_c(T)\, \varphi_o = (F_1 T + F_o)\, \varphi_o \\ K_o(T) : &= F_c(T) + \alpha = F_1 T + F_o + \alpha \end{aligned} \quad . \tag{6.8}$$

We recall that both coefficient functions K_o and K_2 are involved in the definition of the function Φ.

As a result the viscoplastic potential F defined by (6.2) depends on stress invariants and on temperature as well: $F(\sigma,\overline{\sigma},T) := \Phi(\sigma,\overline{\sigma},T) + g(\overline{\sigma})$.

For the determination of the functional dependency of the viscosity coefficient k on temperature one has used creep tests performed at various temperatures. Using these data one has determined for each of the six chosen temperatures the irreversible stress work per unit volume at the beginning and at the end of each creep period taking place at various constant stress levels considered. The viscosity coefficient follows from the already mentioned formula

$$k(\sigma,\overline{\sigma}) = \ln\left|\frac{H(\sigma,\overline{\sigma}) - W(t_i)}{H(\sigma,\overline{\sigma}) - W(t_f)}\right| \frac{H(\sigma,\overline{\sigma})}{\dfrac{\partial F}{\partial \sigma_1}\sigma_1} \frac{1}{t_f - t_i} \tag{6.9}$$

applied for each constant stress level and for each temperature level considered. Here t_i and t_f are the "initial" and the "final" moments of time interval in which creep takes place under

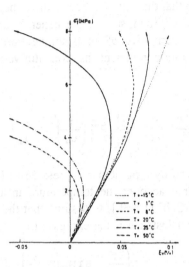

Fig.6.4 Temperature influence on volumetric compressibility and dilatancy of bituminous concrete in uniaxial unconfined tests according to model prediction.

constant stress and constant temperature. As a result, a linear dependency on temperature was obtained

$$k(T) = k_1 T + k_o \qquad (6.10)$$

with $k_1 = 1.554$ (°C)$^{-1}$ and $k_o = 52.36$. The relationship (6.10) is a preliminary obtained result. More tests are necessary to establish this relationship based on a greater number of data.

Thus the whole constitutive equation for transient creep which may depend also on temperature is entirely determined. We can now show how this constitutive equation predicts the temperature influence on the stress-strain curves obtained in uniaxial unconfined tests, performed at various temperatures. The way in which temperature influences compressibility and dilatancy according to model prediction is shown in Fig.6.4. While qualitatively the model describes quite reasonable the features of temperature influence, from quantitative point of view more works is necessary in order to establish the constitutive coefficients on broader experimental base.

6.2 Stationary creep

The temperature influence on stationary creep is a much simpler problem, but at the same time an important one, since stationary creep is acting generally during long intervals of time which means also that the consequence of the temperature influence will have time to build up. Let us consider, for instance, the general constitutive equation (4.14). It is assumed that mainly the derivative $\partial S/\partial \tau$ is temperature sensitive via the constitutive function $Q'(\tau)$. For one strain component this constitutive equation in the dilatant domain is (see (4.16)):

$$\dot{\epsilon}_1^I = \frac{b}{3}\left(\frac{\tau}{\sigma_\star}\right)^m\left(\frac{\sigma}{\sigma_\star}\right)^n + \frac{2\sigma_1-\sigma_2-\sigma_3}{9\tau}\left[\frac{bm}{n+1}\left(\frac{\tau}{\sigma_\star}\right)^{m-1}\left(\frac{\sigma}{\sigma_\star}\right)^{n+1} + Q'\left(\frac{\tau}{\sigma_\star}\right)\right] \qquad (6.11)$$

with $Q'(\tau/\sigma_\star)$ to be replaced by a particular expression as for instance one from (4.21). If we would like to take into account the temperature influence, then it is quite natural (CRISTESCU and HUNSCHE [1993b]) to assume that the coefficients involved in Q' are temperature dependent. For instance if the expression

$$p\left(\frac{\tau}{\sigma_\star}\right)^r \sinh\left(q\frac{\tau}{\sigma_\star}\right) \qquad (6.12)$$

is used for $Q'(\tau/\sigma_\star)$ then it will be assumed that the coefficients p and q are temperature dependent and this expression is to be written as

$$p_1\exp\left(-\frac{\bar{Q}}{RT}\right)\left(\frac{\tau}{\sigma_\star}\right)^r \sinh\left(q_1\,q_2^T\frac{\tau}{\sigma_\star}\right) \qquad (6.13)$$

where p_1, q_1, and q_2 are constants, $\bar{Q}$ is the activation energy, R - the universal gas constant

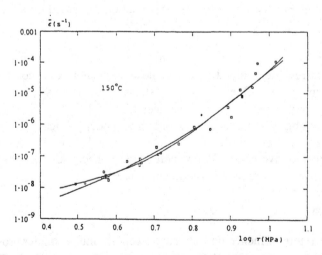

Fig.6.5 Equivalent rate of strain versus τ; model prediction compared with data.

(R= 8.314 J/mol K) and T the absolute temperature. A similar procedure is used for any other expression used for Q' (see for instance (4.21)). The constitutive coefficients can be determined using stationary creep tests performed at various temperatures. The results are matching very well the data (see HUNSCHE and CRISTESCU [1994]- in preparation). As an example in Fig.6.5 (CRISTESCU and HUNSCHE [1993b]) are shown the experimental data and the model prediction for creep of rock salt at temperature 150°C. Formula (4.20) is used for this purpose.

Also for rock salt and stationary creep WANTEN et al.[1993] use a uniaxial formula of the form

$$\dot{\epsilon} = A \sigma^n \exp\left(-\frac{\bar{Q}}{RT}\right) \qquad (6.14)$$

where n=12 and $\bar{Q}$=105 ± 20 kJ/mole. The same type of law (6.14) has been used by HUNSCHE et al.[1993] where, however, the stress is replaced by the difference between the applied stress and the internal back-stress.

7. MINING ENGINEERING EXAMPLES

In order to illustrate how the constitutive equations given in the previous sections can be used, some examples will be given. We will choose the simplest examples. We start with a mathematical formulation of the problem (see also CRISTESCU [1989a], Chap.11).

7.1 Formulation of the problem.

We call the stresses existing in situ before excavation "primary" stresses and the notation $\sigma^P(X)$ will be used. These vary from one location X to another one, but are constant in time. We make the standard assumption according to which σ^P depends only on the depth h, and that the vertical component σ_V is due to the overburden pressure

$$\sigma_V(h) = \int_0^h \rho \, z \, dz = \gamma h = 0.025h \qquad (7.1)$$

with h in meters and σ_V in MPa. What concerns the horizontal components, we denote by σ_h the mean horizontal component and will consider several cases obtained from $\sigma_h = n \, \sigma_V$ with n a constant comprised in the interval $0.3 \le n \le 1.5$ and only seldom smaller or bigger; its value depends on the location (see CRISTESCU [1989a] Chap.10, MARTINETTI and RIBACCHI [1980], RUMMEL and BAUMGARTNER [1985], HERGET [1987], COOLING et al.[1988], CUNHA [1990]). With the above assumptions σ^P satisfies the Cauchy balance of linear momentum equations.

Further we denote by $\sigma^S(X,t)$ the "secondary" stresses, i.e. the stresses after excavation and by $\sigma^R(X,t) = \sigma^S(X,t) - \sigma^P(X)$ the "relative" stresses.

The constitutive equation

$$\dot{\varepsilon} = \frac{\dot{\sigma}}{2G} + \left(\frac{1}{3K} - \frac{1}{2G}\right)\dot{\sigma}\,\mathbf{1} + k_T\left(1 - \frac{W(t)}{H(\sigma)}\right)\frac{\partial F}{\partial \sigma} + k_S\frac{\partial S}{\partial \sigma} \qquad (7.2)$$

must be satisfied by the primary stress-state when $\dot{\sigma} = 0$. Therefore before excavation we must have

$$H(\sigma^P) = W^P = const. \qquad and \qquad \dot{\varepsilon}^P = k_S\frac{\partial S(\sigma^P)}{\partial \sigma} \ . \qquad (7.3)$$

The relation $(7.3)_2$ describes the slow tectonic motions if $\dot{\sigma}^P$ has a significant value. For primary stress states close to the hydrostatic one, when $\sigma_h \approx \sigma_v$, the rate of deformation components obtained from $(7.3)_2$ are negligible with respect to the motion taking place after excavation. Condition $(7.3)_1$ represents a condition of "stabilization" or "equilibrium" for the transient creep. If due to a fast erosion or excessive previous excavations above the location considered for the future excavation we have $H(\sigma^P) < W^P$, then the problem is slightly more involved. Here we assume $(7.3)_1$ everywhere in the primary state.

In order to obtain a very simple solution we assume that the excavation is done quite fast with respect to the time-span during which the excavation (tunnel, shaft, etc.) is in use. Thus the initial stress variation due to the excavation will be assumed to be elastic, i.e. "instantaneous". Afterwards, in the long time interval which follows, the geomaterial surrounding the excavation will deform by creep, possibly associated with a stress relaxation. The strains produced by the excavation will be related to the state existing *in situ* before excavation considered to be the reference configuration. Thus the "relative" stress components are responsible for the strains produced by the excavation.

The procedure we are using is the following. We assume that $\sigma^P(X)$ are measured and therefore known at the location of excavation. Assuming a fast excavation, we determine the secondary stress $\sigma^S(X,t)$ and therefore also the relative one $\sigma^R(X,t)$. Introducing this stress variation in the constitutive equation we determine the relative strains $\varepsilon^R(X,t)$ and further the relative displacements $u^R(X,t)$. Let us observe that what we can measure after excavations are just secondary stresses, relative strains and relative displacements. Preliminary measurements of primary stresses are necessary. In order to be correct, a solution must satisfy the following obvious requirements: if the distance from the wall of excavation increases very much, then:

$$\lim_{dist\to\infty} \boldsymbol{u}^{R}=\boldsymbol{0} \quad , \quad \lim_{dist\to\infty} \boldsymbol{\epsilon}^{R}=\boldsymbol{0} \quad , \quad \lim_{dist\to\infty} \boldsymbol{\sigma}^{R}=\boldsymbol{0} \qquad (7.4)$$

i.e. far enough from the wall of the excavations the displacements and strains must be zero while the stresses must approach their primary values (the "far field stresses"). In most cases the initial secondary stresses are obtained as elastic solutions.

Let us discuss the formulation of the boundary conditions. The boundary conditions at infinity are just (7.4). What concerns the condition at the surface of the excavation, several cases can be considered. If the walls of the excavation are not supported, we must express the condition that the stress vector acting on the wall has zero components (stress-free surface), i.e.

$$(\boldsymbol{\sigma}^{S})^{T}\boldsymbol{n}=\boldsymbol{0} \qquad (7.5)$$

where the superscrpt T stands for "transpose" and n is the normal to the surface and oriented towards the interior of the opening. If the opening is filled with a fluid or gas under pressure, then a pressure is exerted on the walls of the cavity

$$(\boldsymbol{\sigma}^{S})^{T}\boldsymbol{n}=p\boldsymbol{n} \qquad (7.6)$$

where p is the pressure. This pressure depends on the depth in the case of a fluid (weight of the column). If the cavern is filled with a gas, we have considered two cases: the gas is under a constant controlled pressure p_o, say, or the pressure of the gas in a "closed" cavern is increasing due to the creep closure of the walls . Thus for the pressure involved in (7.6) we can consider three variants:

$$p=\gamma_1 h \quad , \quad p=p_o=const. \quad , \quad pV=const. \qquad (7.7)$$

where V is the volume of the cavern.

The case when on the walls of the underground excavation are acting some forces due to a man-made support, will not be discussed here. For a theory of rigid, elastic and yieldable supports see CRISTESCU [1989a] Chap.13.

7.2 Creep around a vertical shaft or borehole.

One of the simplest possible problems is the study of creep, dilatancy and/or compressibility, damage, failure and creep failure around a cylindrical vertical cavity, either a vertical shaft or a borehole, or a cavern used for the storage of petroleum products, etc. (we follow here CRISTESCU [1985a,b],[1989a, Chap.11],[1992]; see also FISCHER *et al.*[1992]).

It will be assumed that the cavity is quite deep so that the influence of the free surface will be disregarded.Thus the shaft, say, is a circular cylindrical orifice drilled in a semi-infinite space. Cylindrical coordinates r, θ, and z will be used, with the z-axis coinciding with the symmetry axis of the cavity and directed vertically downwards. The initial radius is r=a. The problem is assumed to be a "plane strain" problem and the solution will be given for a fixed depth h. Thus the boundary conditions (7.6) at r=a and a fixed depth h are:

$$r = a \quad , \quad z = h : \quad \sigma_s^R = p - \sigma_h \quad , \quad \sigma_{rz}^R = \sigma_{r\theta}^R = 0 \quad . \qquad (7.8)$$

Here the depth is involved via the particular constant values of σ_v and σ_h, maybe in p too.

Due to the symmetry of the problem, the strain components in cylindrical coordinates are:

$$\epsilon_r^R = \frac{\partial u^R}{\partial r} \quad , \quad \epsilon_\theta^R = \frac{u^R}{r} \quad , \quad \epsilon_z^R = 0 \qquad (7.9)$$

while the equilibrium equation and compatibility equation are

$$\frac{\partial \sigma_r^R}{\partial r} + \frac{\sigma_r^R - \sigma_\theta^R}{r} = 0 \quad , \quad \frac{\partial \epsilon_\theta^R}{\partial r} + \frac{\epsilon_\theta^R - \epsilon_r^R}{r} = 0 \quad . \qquad (7.10)$$

The elastic solution is obtained quite easy (see for instance CRISTESCU [1985a,b],[1989a] Chap.11):

$$\sigma_r^S = (p - \sigma_H) \frac{a^2}{r^2} + \sigma_h \qquad\qquad \epsilon_r^R = \frac{p - \sigma_h}{2G} \frac{a^2}{r^2}$$

$$\sigma_\theta^S = -(p - \sigma_H) \frac{a^2}{r^2} + \sigma_h \qquad\qquad \epsilon_\theta^R = -\frac{p - \sigma_h}{2G} \frac{a^2}{r^2} \qquad (7.11)$$

$$\sigma_z^S = \sigma_v \qquad\qquad\qquad\qquad \epsilon_z^R = 0$$

$$\sigma^S = \sigma^P$$

$$w^R = 0 \qquad\qquad\qquad\qquad u^R = -\frac{p - \sigma_h}{2G} \frac{a^2}{r}$$

If the excavation or drilling is performed very fast, then the stresses, strains and displacements obtained from (7.11) are just those existing after excavation. Afterwards the rock will deform by creep, in some domains around the cavity being dilatant while in some others compressible. The boundaries of these domains can be determined quite easy if we introduce the stresses (7.11) in the condition (2.17) to see where a loading occurs and afterwards in the conditions (2.18)-(2.20) to further determine where a compressibility takes place during loading, and where a dilatancy. There where inequality (2.21) holds an unloading will take place.

In Fig.7.1 (CRISTESCU and HUNSCHE [1993c]) is shown the stress distribution along the walls of a deep cavern excavated in rock salt as function of depth. The dash-dot

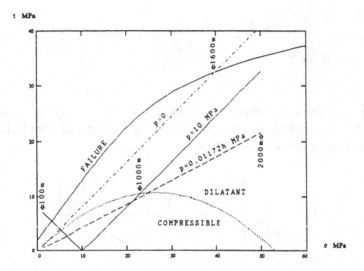

Fig.7.1 Secondary stress states in the walls of a deep well: empty well (dash-dot line),well filled with brine (interrupted line) and well filled with gas under pressure (full line)

line corresponds to the empty well; in this case instantaneous failure will occur at depths greater than 1600 m; up to this depth the stress states at the well wall are in the dilatancy domain. For depths approaching 1600 m, one can expect a creep failure to take place after a certain time period. If the well is filled with brine the stress states along its walls at various depths is shown by an interrupted line. In this case it is only for depths greater than approximately 1000 m that the rock surrounding the well is dilatant, while up to this depth it is compressible. If the well is filled with a gas under constant pressure (10 MPa, say) then the stress at various depths is shown as full line. In this case at great depths the surrounding rock is dilatant, at medium depths is compressible, while at small depths again a dilatant region is possible, followed at shallow depths by failure. In the case of closed caverns, the stress distribution is similar to the last one (constant pressure) but as the pressure is building up due to the convergence of the walls, the octahedral shear stresses at shallow depths are increasing, while at greater depths are decreasing.

For the study of convergence by creep of the walls of the cavern we can use the formulae presented in the previous sections.The biggest part of the convergence is due to the stationary creep (see also the remarks by FRAYNE et al.[1993]). For instance, a simple illustrative solution can be obtained if we assume that during creep the stress state remains constant. In this case we obtain from (4.36) for the radial displacement

$$\frac{u}{a} = -\frac{p-\sigma_h}{2G}\frac{a}{r} +$$

$$\frac{r}{a}\left\{\frac{\left\langle1-\frac{W_T(t_o)}{H}\right\rangle\frac{\partial F}{\partial\sigma_\theta}}{\frac{1}{H}\frac{\partial F}{\partial\sigma}\cdot\sigma}\left\{1-\exp\left[\frac{k_T}{H}\frac{\partial F}{\partial\sigma}\cdot\sigma(t_o-t)\right]\right\}+k_s\frac{\partial S}{\partial\sigma_\theta}(t-t_o)\right\} \qquad (7.12)$$

where t_o is the moment of excavation. The depth is involved in (7.12) via σ_h and the stress state. The time t_c of complete closure of the cavern can be obtained from (7.12) when for r=a the displacement u becomes equal to a:

$$t_c \approx t_o + \frac{1}{k_T\frac{\partial S}{\partial\sigma_\theta}}\left\{1+\frac{p-\sigma_h}{2G}-\frac{\left\langle1-\frac{W_T(t_o)}{H}\right\rangle\frac{\partial F}{\partial\sigma_\theta}}{\frac{1}{H}\frac{\partial F}{\partial\sigma}\cdot\sigma}\right\} . \qquad (7.13)$$

Here the last right hand term is the ultimate value (for t →∞) of the transient component of the displacement. Thus in the time interval $t_s<t<t_c$ the closure is entirely due to the stationary creep.

In the dilatancy domain a progressive damage takes place during creep. For stress states which are close to the surface of instantaneous failure the damage process is sped up. We can estimate the amount of damage by using a formula of the type (5.5) and creep failure using (5.3) in conjunction with (5.5). Also the spreading in time into the rock mass of the dilatancy can be determined with the last two terms from the formula (4.40). Let us give an example.

In Fig.7.2 (left) are shown the successive positions of the walls of an empty (p=0) vertical cavern excavated in rock salt, according to (7.12). At great depths the closure is very fast, but it slows down exponentially as the depth diminishes, i.e. at shallow depths very long time intervals are necessary even for a partially closure of the cavern. Apparently it looks like the bottom is raising, but the present solution is obtained considering radial convergence only (see also MUNSON et al.[1992]). In Fig.7.2 the five cases shown correspond to five values attributed to the dimensionless time variable $k_s(t-t_o)$ ($t_1=4\times10^4$,

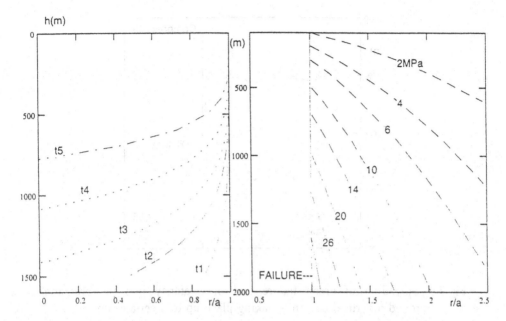

Fig.7.2 Successive positions and closure due to creep of the wall of a vertical
cavern excavated in rock salt (left) and distribution of octahedral shear
stress τ just after excavation, explaining why at greater depths the closure
is much faster (right).

$t_2=1.6\times10^5$, $t_3=4\times10^5$, $t_4=1.52\times10^6$ and $t_5=8\times10^6$). The correct value of k_s needs further
studies, but it is certainly the parameter which influences the most the speed of
convergence, besides the depth. For instance, choosing $k_s= 3\times10^{-3}$ s^{-1}, the five cases shown
correspond to 154 days, 1.7 year, 4.2 y, 16 y, and 84 years, respectively.

The reason for the much faster closure at great depths is shown in Fig.7.2 (right):
according to the elastic solution the curves τ=const. are looking as shown in this figure.
Therefore, at great depths and mainly close to the wells of the cavern the values of τ are
quite big. For depths greater than 1600 m, say, an instantaneous failure is to be expected,
as shown in the figure by a solid line. Let us recall also that all stationary creep laws are
expressed as power functions of τ. That is why there where we have big values of τ, the
magnitude of stationary rate of strain may become huge. For instance, for rock salt (m=5
in (4.8) and $q_o=5$ in (4.21) and (4.15)) for the values of τ shown in Fig.7.2 (right), the
magnitude of the stationary rate of strain is ranging between 1 and 10^7. That is why at great
depths the closure is much faster, while at shallow depths the closure may be even
negligible.

Let us give an example of the spreading of creep failure and of dilatancy during

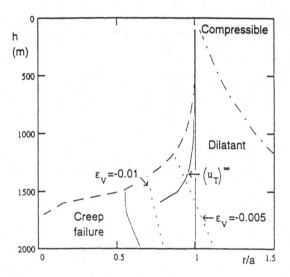

Fig.7.3 Spreading of creep failure and of dilatancy into the creeping rock around a vertical cavern as taking place up to a certain time.

wall closure. In Fig.7.3 is shown the position of the converging walls at a certain arbitrary chosen time (interrupted line). The vertical cavern is excavated in rock salt and it is empty (p=0). At the initial moment the walls of the cavern are r=a, while the initial compressibility/dilatancy boundary is shown as a dash dot line. However, at far distances the dilatancy is quite small, the rock being practically incompressible. Also instantaneous failure takes place just after excavation at depths greater than 1600 m (see also Fig.7.1). The initial shape of the failed domain is not shown, but it is shown in Fig.7.2 (right); its ultimate location at the moment of full closure is evident by the slope discontinuity of the dotted line. The boundary of the domain where the creep failure has already taken place at the time considered, is shown as a full line. Two dotted lines show how dilatancy progresses into the creeping rock. A full line shows the ultimate position of the walls due to transient creep alone, according to formula (4.38). It is obvious that transient creep would not be able to describe a full closure of a cavern, and the main contribution is that of the stationary creep.

Let us give another example of a closed cavern, i.e. of a cavern filled with gas under pressure and the creeping rock increases this pressure according to a law of the form pV = const., say. Any other law could equally be used. In Fig.7.4 is shown an example. The lower interrupted line was taken as "initial" shape of the cavern when the internal pressure was p=5 MPa, say. At that moment the cavern is closed and the pressure is building up due to the creep of rock. The upper interrupted line shows the subsequent position of the walls when the internal pressure has increased up to 5.928 MPa. Due to the pressure built up the dilatancy/ compressibility boundary is also changing as well as the

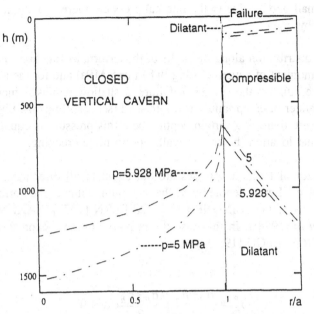

Fig.7.4 Convergence of the walls of a closed cavern, and dilatancy and compressibility around the cavern.

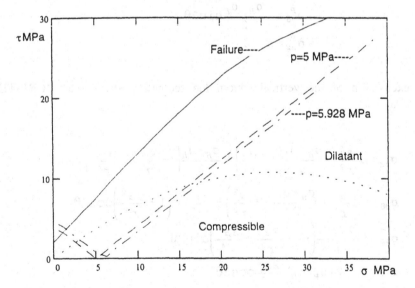

Fig.7.5 Variation of pressure in a closed cavern.

failure domain at shallow depths. This failure has no significance anyway since no real cavern under pressure is excavated up to shallow depths. It is, however, interesting to mention that at small depths, due to the internal pressure, there is a slight tendency of the cavern to expand by creep.

The stress distribution along the walls of the cavern, as function of depth, is shown in Fig.7.5 for the moment of cavern closing (when p=5 MPa) and for the actual time. Thus due to the pressure built up the danger of failure at shallow depths is increased, while at great depths the danger of creep failure is diminished (see also Fig.7.4). Fig.7.5 shows also that for each pressure there is a certain depth where this pressure is equal to the far field stress, so that at that location the cavern walls are no more creeping.

If the horizontal far field stresses are not equal in all directions, the problem is slightly more involved. Let us denote by σ_H the maximum value of the horizontal stress and by σ_h the minimum one (see LINDNER and HALPERN [1978], SAXENA et al.[1979], ABOU SAYED et al.[1978]). In this case the primary stresses expressed in cylindrical coordinates are (CRISTESCU [1985a]):

$$\sigma_{rr}^P = \frac{\sigma_H + \sigma_h}{2} + \frac{\sigma_H - \sigma_h}{2}\cos 2\theta$$

$$\sigma_{\theta\theta}^P = \frac{\sigma_H + \sigma_h}{2} - \frac{\sigma_H - \sigma_h}{2}\cos 2\theta$$

$$\sigma_{r\theta}^P = -\frac{\sigma_H - \sigma_h}{2}\sin 2\theta \qquad\qquad (7.14)$$

$$\sigma_{zz}^P = \sigma_V \;.$$

After the excavation of the vertical cavern the secondary stresses are (CRISTESCU [1985c][1989a]):

$$\sigma_{rr} = p\frac{a^2}{r^2} + \frac{\sigma_H + \sigma_h}{2}\left(1 - \frac{a^2}{r^2}\right) + \frac{\sigma_H - \sigma_h}{2}\left(1 - \frac{4a^2}{r^2} + \frac{3a^4}{r^4}\right)\cos 2\theta$$

$$\sigma_{\theta\theta} = -p\frac{a^2}{r^2} + \frac{\sigma_H + \sigma_h}{2}\left(1 + \frac{a^2}{r^2}\right) - \frac{\sigma_H - \sigma_h}{2}\left(1 + \frac{3a^4}{r^4}\right)\cos 2\theta$$

$$\sigma_{r\theta} = -\frac{\sigma_H - \sigma_h}{2}\left(-1 - \frac{2a^2}{r^2} + \frac{3a^4}{r^4}\right)\sin 2\theta \qquad\qquad (7.15)$$

$$\sigma_{zz} = \sigma_V - \nu(\sigma_H - \sigma_h)\frac{2a^2}{r^2}\cos 2\theta$$

Now the stress distribution around the circumference of the cavern depends on orientation (on θ). For instance it we assume a primary stress distribution of the form

$$\sigma_V = 0.027h, \quad \sigma_h = 5.4 + 0.012h, \quad \sigma_H = n\sigma_h$$

with n= 1, 2, 4 and 6, the stress distribution is shown in Fig.7.6 for a vertical borehole drilled in granite at h= 2 km and p=0. The diamonds correspond to n=1: full diamond - primary stress, empty diamond - secondary stress. Similarly, the circles correspond to n=2,

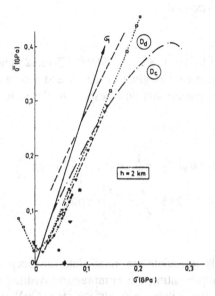

Fig.7.6 Stress state around a borehole drilled in granite for various ratios σ_H/σ_h.

triangles to n=4 and squares to n=6. Tensile failure is possible in the direction of σ_H and failure by dilatancy in the direction of σ_h. In Fig.7.6 the interrupted line is the failure line, while the dash-dot line the compressibility/dilatancy boundary. Thus, if $\sigma_H \neq \sigma_h$ around the circumference we can have domains of dilatancy and some other domains of compressibility, and the problem must be analyzed at each depth. This problem is similar with the one encountered with horizontal tunnels, and which will be discussed in detail below.

7.3 Creep around a horizontal tunnel.

Let us assume that the circular tunnel is located at a certain depth where the values of the primary horizontal and vertical stresses σ_h and σ_v, generally distinct, are known. Since their values vary smoothly with the depth, and since the radius of a deep tunnel is much smaller than the depth of the location, we assume that in the immediate neighborhood of the tunnel these components are practically constant and equal to their values at the depth of the tunnel axis. Using cylindical coordinates and assuming plane strain state one obtains quite easy the stress state just after a fast excavation:

$$\sigma_{rr}^S = p\frac{a^2}{r^2} + \frac{1}{2}(\sigma_h + \sigma_v)(1 - \frac{a^2}{r^2}) + \frac{1}{2}(\sigma_h - \sigma_v)(1 - \frac{4a^2}{r^2} + \frac{3a^4}{r^4})\cos2\theta$$

$$\sigma_{\theta\theta}^S = -p\frac{a^2}{r^2} + \frac{1}{2}(\sigma_h + \sigma_v)(1 + \frac{a^2}{r^2}) - \frac{1}{2}(\sigma_h - \sigma_v)(1 + \frac{3a^4}{r^4})\cos2\theta$$

$$\sigma_{r\theta}^S = \frac{1}{2}(\sigma_h - \sigma_v)(-1 - \frac{2a^2}{r^2} + \frac{3a^4}{r^4})\sin2\theta \qquad (7.16)$$

$$\sigma_{zz}^S = \sigma_h - \nu(\sigma_h - \sigma_v)\frac{2a^2}{r^2}\cos2\theta$$

(see, for instance CRISTESCU [1985a,b],[1989a] Chap.12 where a derivation of these formulae for the plane strain case is given in detail and where formulae for strains were also given).The relative displacements with respect to the state in situ before excavation are

$$u_r^R = \frac{1+\nu}{E}\left\{-p\frac{a^2}{r} + \frac{1}{2}(\sigma_h + \sigma_v)\frac{a^2}{r} + \frac{1}{2}(\sigma_h - \sigma_v)\left[4(1-\nu)\frac{a^2}{r} - \frac{a^4}{r^3}\right]\cos2\theta\right\}$$

$$u_\theta^R = -\frac{1+\nu}{E}\frac{1}{2}(\sigma_H - \sigma_v)\left[2(1-2\nu)\frac{a^2}{r} + \frac{a^4}{r^3}\right]\sin2\theta \quad . \qquad (7.17)$$

In order to correctly formulate the initial data for creep it is useful to illustrate in a $\sigma\tau$ - plane the curves representing the compressibility/dilatancy boundary (X=0), the failure surface (Y=0) and the initial yield surface (H(σ,τ)=W^P), together with the point representing the primary stress-state (σ^P,τ^P) and the points representing the secondary stress state (7.16). This plane is somehow a "map" of the model and initial stress state. In fact we do not need the whole constitutive equation in order to plot this "map"; we need only the expressions of X(σ,τ) and Y(σ,τ) which yield straightforward from triaxial tests as well as the yield function H(σ,τ), which is certainly more difficult to be found.

An example of such a map is shown in Fig.7.7 for a tunnel excavated in rock salt at h=1000 m, the ratio of far field stresses being σ_h/σ_v=0.3 and no internal pressure (p=0) (CRISTESCU and HUNSCHE [1993c]). The diamond represents the primary stress. One dotted line represents the secondary stress distribution along the circumference r=a, from θ=0° up to θ=90° (roof). If the ratio σ_h/σ_v would be closer to unity, then the length of this segment would be smaller.The other two dotted lines represent the secondary stress along the circumferences r=2a and r=3a respectively. Thus, as the distance from the walls of the tunnel increases the secondary stress states come closer to the far field stresses. The initial yield surface is shown as an interrupted line. Fig.7.7 also shows that instantaneous failure occurs on the tunnel circumference both around θ=0° and θ=90°. Certainly that creep failure is to be expected close to θ=0°, there where high values of τ are involved; this creep failure will afterwards spread into the rock mass. That will be discussed below. If the order

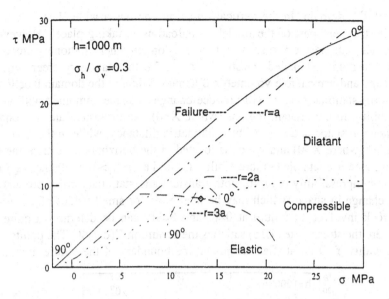

Fig.7.7 Map showing the secondary stress distribution just after excavation of a horizontal tunnel for $\sigma_h/\sigma_v=0.3$ and $p=0$.

of magnitude of the far field stresses is interchanged, i.e. $\sigma_h<\sigma_v$, then the creep failure is expected at $\theta=90°$, etc. Furthermore, Fig.7.7 shows that around the circumference the stress state can be located in various loading or unloading domains. This is shown also in Fig.7.8 which is the "map" of the tunnel and the surrounding rock (CRISTESCU [1985a,b]). A wide region of dilatancy (where $X<0$) is located in the horizontal direction with an island

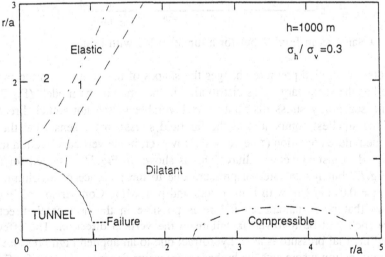

Fig.7.8 Domains of compressibility, dilatancy, elasticity and failure around a horizontal tunnel excavated in rock salt ($p=0$, $\sigma_h/\sigma_v=0.3$).

of compressibility domain. In the vertical direction one has a wide elastic domain (with respect to the transient part of the model an unloading is taking place). However, if the model describes stationary creep as well, this is a domain where stationary creep holds as follows: in the domain labelled "elastic 1" in Fig.7.8 the stationary creep will produce change in shape and irreversible volumetric dilatancy, while in the domain labelled "elastic 2" (where X>0) stationary creep will produce change in shape combined with volumetric incompressibility. In the dilatant regions (where X<0), both transient and stationary creep produce change in shape and irreversible volumetric dilatancy, while in the island labelled "compressible" (where X>0) and which is located in the horizontal direction, the transient creep will produce irreversible compressibility (X>0, i.e. $(\dot\varepsilon_v^I)_T>0$) while the stationary one volumetric incompressibility $(\dot\varepsilon_v^I)_S=0$; both transient and stationary creep produce here an irreversible change in shape (which may be negligible for small values of τ). Finally, the amount of rock involved in failure at $\theta=0°$ and $\theta=90°$ can be determined quite easy by checking where the stress state (7.16) satisfies the condition $Y(\sigma,\tau)\le0$. The points in Fig.7.8 where the equality $Y=0$ is satisfied are just at the boundary of the failure domain.

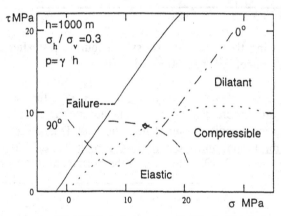

Fig.7.9 Same as in Fig.7.7 but for a tunnel filled with brine.

An internal applied pressure changes the shapes of the domains on the map of the tunnel as well as the secondary stress distribution in the "map of the model" (Fig.7.7). This change in the secondary stress distribution is favorable in the horizontal direction (the direction of the smallest component of the far field stress) in the sense that the stresses depart from the failure condition (τ decreases). However, in the vertical direction more rock will be involved in instantaneous faliure. This is shown in Fig.7.9 obtained for the same case as in Fig.7.7 but for an internal pressure due to the presence of a column of brine according to p= 0.011772 h, with h in meters and p in MPa. Comparison of Fig.7.9 and Fig.7.7 shows that no instantaneous failure is possible in the horizontal direction, but instead more rock will be involved in failure in the vertical direction. The effect of the superimposed internal pressure is as in hydrofrac: due to an applied high pressure the rock will fail in the direction where acts the higher compressive component of the far field stress (CRISTESCU [1989a] Chap.12). The various domains existing around a tunnel for the same case as in Fig.7.9 are shown in Fig.7.10. The stress states existing around the

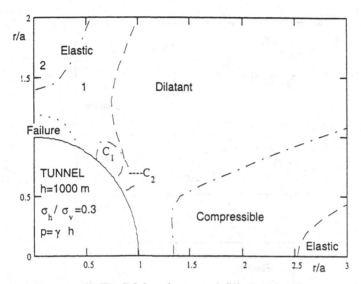

Fig.7.10 Same as in Fig.7.8 but for tunnel filled with brine.

circumference can be followed easier if one compares Fig.7.10 with Fig.7.9. Again "elastic" refers to transient creep, and stationary creep is present in all domains. C_1 is a compressibility domain for transient creep, while C_2 is an elastic domain for transient creep. Both C_1 and C_2 are domains of incompressibility for stationary creep. "Elastic 1" is a dilatant region for stationary creep, while "elastic 2" is an incompressible domain for stationary creep.

Creep failure by dilatancy is quite possible in the case shown in Fig.7.7 since in the neighborhood of the point r=a, $\theta=0°$ the octahedral shear stress is significantly big. However, if the internal applied pressure is high, then the possibility of a creep failure is significatly curtailed (see for further details CRISTESCU [1986], [1989a],[1993c]).

The examples given above are somehow extreme in the sense that the chosen ratio σ_h/σ_v is considerably small, but possible. It was chosen in this way to show how non-symmetric the domains of dilatancy, compressibility and elasticity can be, and how far can spread the dilatancy domain in this case. Let us recall that dilatancy means increase of permeability (HUNSCHE and SCHULZE [1993]) with all consequences for a design of a radioactive repository or for a hazardous chemical waste, for instance.

Let us give some examples when the ratio σ_h/σ_v is somehow bigger. For instance, in the lower part of the Fig.7.11 are shown the domains of dilatancy, compressibility and elasticity around a tunnel excavated in rock salt at the depth h=1000 m, for the ratio σ_h/σ_v=0.5 and p=0. For the same case, in the upper part of the same figure is shown the stress distribution along the contour. In this case no instantaneous failure is present in the

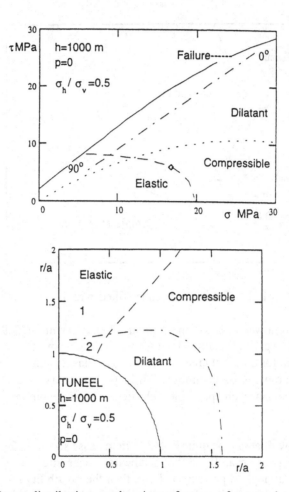

Fig.7.11 Stress distribution on the circumference of a tunnel and domains of significance around the tunnel (h=1000 m, σ_h/σ_v=0.5, p=0)

horizontal direction but in the same direction a creep failure is to be expected after a certain time since the values of τ are high. Again, that is the direction of the smallest component of the far field stress. The presence of an internal pressure significantly reduces the danger not only of instantaneous failure but also that of the creep failure in the horizontal direction as can be seen in Fig.7.12. The octahedral shear stress at $\theta=0°$ decreases significantly, while the increase of τ at $\theta=90°$ is not yet so important as to produce instantaneous failure in that direction; a much higher internal pressure would be necessary.

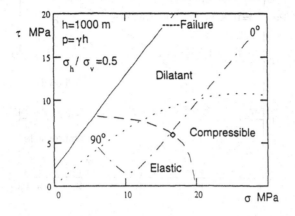

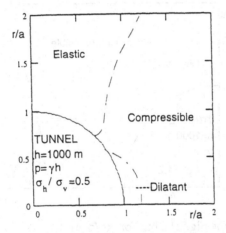

Fug.7.12 Same as for Fig.7.11 but with an internal pressure due to a brine column.

If the ratio σ_h/σ_v is close to unity, then the regions of dilatancy are circular around the tunnel opening. This last case is a very simple one, since dilatancy and tunnel closure can be described by the formulae (4.40) and (7.12). Let us give an example when the ratio σ_h/σ_v is close to unity, but somehow higher, i.e. $\sigma_h/\sigma_v=1.2$. In this case (Fig.7.13) the domain of dilatancy is still close to a circular one with a slight enlargement in the vertical direction. In the same direction we find now the maximum value of τ and the minimum far field stress. Instantaneous failure is not possible but creep failure is possible after a long time interval, in the vertical direction in the neighborhood of the point r=a, $\theta=90°$, there where τ is maximum.

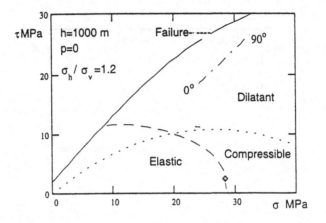

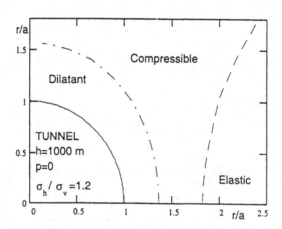

Fig.7.13 Same as for Fig.7.11 but for $\sigma_h/\sigma_v=1.2$.

If the ratio σ_h/σ_v is still higher, an instantaneous failure is possible in the vertical direction. This is shown in Fig.7.14 for the ratio $\sigma_h/\sigma_v=2$. One can see that a significant portion of the rock in the vertical direction (roof) will fail instantaneously just after excavation. The amount of rock involved in such failure can be determined very easy by checking where the stress state (7.16) satisfies the inequality $Y(\sigma,\tau)\leq0$. This is shown in Fig.7.15 which corresponds to the same case as in Fig.7.14. The dotted line shows the boundary of the failed domain. Besides the instantaneous failure occurring between $\theta=90°$ and $\theta=49.5°$, just outside the boundary of the failed region (where equality $Y=0$ holds) a creep failure is also to be expected there where the value of τ is very high and close to the failure surface. Thus in the case shown in Fig.7.14, instantaneous failure is followed by

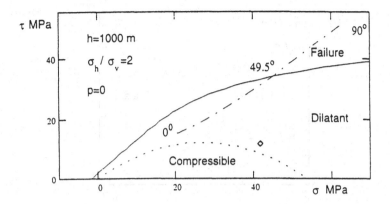

Fig.7.14 Secondary stress distribution just after excavation, in the rock surrounding the tunnel for $\sigma_h/\sigma_v=2$.

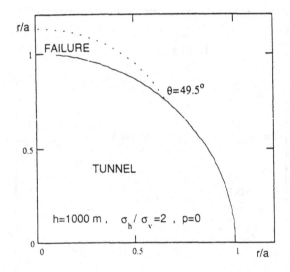

Fig.7.15 Domain at the roof where instantaneous failure will take place just after excavation (same case as in Fig.7.14).

creep failure in the same direction where is acting the minimum far field stress.

From all the cases discussed above we can conclude that instantaneous failure may occur either in the direction σ^P_{max} or σ^P_{min}, if one of the failure conditions $Y \leq 0$ or $\sigma_i \leq -\sigma_t$ (i=1, 2, 3 and σ_t the tensile strength of the rock) is satisfied; this may happen for small values of τ in the direction of σ^P_{max} if $\sigma_h < \sigma_v$, more likely if internal pressure is present and high; for $\sigma_h < \sigma_v$ it is also possible at high values of τ in the direction of σ^P_{min} (mainly at great depths). However, instantaneous failure is possible in both directions σ^P_{max} and σ^P_{min}

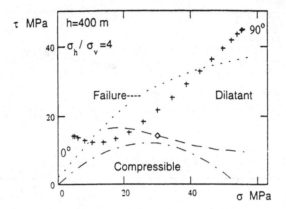

Fig.7.16 Stress distribution around the circumference of a tunnel excavated in rock salt for high value of the ratio σ_h/σ_v.

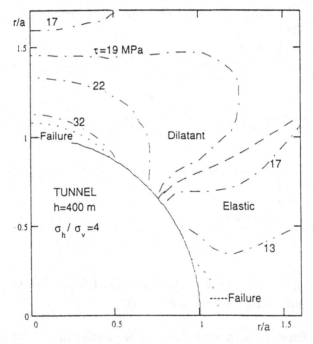

Fig.7.17 Instantaneous failure around a tunnel in the case shown in Fig.7.16.

if $\sigma_h \ll \sigma_v$ or $\sigma_h \gg \sigma_v$. An example for $\sigma_h \gg \sigma_v$ is given in Fig.7.16 for a tunnel excavated in rock salt at h=400 m and ratio σ_h/σ_v=4. In this case both at roof (direction of σ^P_{min}) and wall (direction of σ^P_{max}) failure is due to the condition Y≤0. The difference between the two cases is that at the roof there are high values of τ and therefore after the instantaneous failure may follow in time a creep failure. As already mentioned, creep failure is to be expected there where there are very high values of τ (close to the short term failure surface), always in the direction of σ^P_{min}, either in the case $\sigma_h > \sigma_v$ or $\sigma_h < \sigma_v$. The region of instantaneous failure corresponding to the case in Fig.7.16 is shown as dotted line in Fig.7.17. Various dash-dot lines on the same figure are lines of constant τ. There where τ has high values a significant creep is to be expected. In the case shown in Fig.7.17 much more creep is to be expected in the vertical direction (direction of smallest far field stress).

Let us shortly present now how creep of the rock surrounding the tunnel can be described. For the general case, i.e. arbitrary far field stresses and boundary conditions on the wall of the tunnel, one has to integrate the system of constitutive equation, the equilibrium equation and the compatibility equations. Some particular examples are given by CRISTESCU [1985b][1988] [1989a Cap.11]. However, a very simple approach is possible if the ratio σ_h/σ_v is close to unity and if one assumes that the stresses do not vary

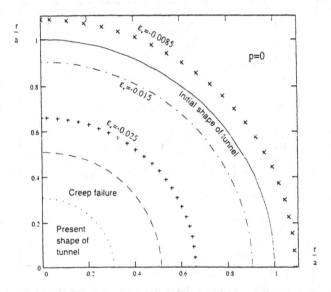

Fig.7.18 Closure of the well of a horizontal tunnel, showing also the spreading of creep failure and dilatancy into the rock mass ($\sigma_h = \sigma_v$, h=700 m).

during creep. This last assumption is certainly somehow restrictive but it can lead to illustrative examples which would obviously describe a much faster creep closure than occurring if the stress relaxationn would also be taken into account. Such solutions have been labelled "simplified" solutions (CRISTESCU [1985b][1989a]). For instance, if $\sigma_h = \sigma_v$ for a depth h=700 m and a tunnel excavated in rock salt with p=0, the creep of the rock surrounding the tunnel is shown in Fig.7.18. The creep closure can be obtained with a formula of the form (7.12). The initial shape of the tunnel, just after excavation is shown as a solid line. The position of the wall of the tunnel after a dimensionless time interval

$k_S(t-t_o)=7.5\times10^6$ is shown as a dotted line. For instance, if we choose $k_S=3\times10^{-3}s^{-1}$ this time interval would correspond to 63 years. The closure rate strongly depends on the value of k_S which has to be determined either in creep tests or just by measuring the convergence rate of the walls of a tunnel. By interrupted line is shown the boundary of creep failure propagation up to the same particular time. The way in which dilatancy spreads into the creeping rock is also shown by three lines ε_v^I=const. Thus the elastic/viscoplastic model allows for the description not only of creep but also of the slow spreading of dilatancy and of creep failure.

If the ratio σ_h/σ_v is distinct from unity, the above formulae cannot be used anymore. The closure will be faster there where the values of τ are higher. For instance, in Fig.7.17 are shown by dash-dot lines the lines of τ=const. for the ratio σ_h/σ_v=4 and h=400 m (see also Fig.7.16). Thus for this case we expect the closure to be faster at the roof in the direction of σ^P_{min}, there where we expect also more dilatancy and maybe creep failure.

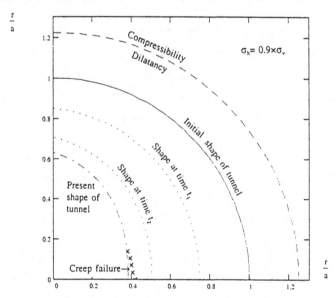

Fig.7.19 Successive positions of the wall of a tunnel for σ_h/σ_v=0.9. Closure is faster in the minor σ^P direction, where τ has high values and creep failure is also expected.

The examples shown in Fig.7.17 and Fig.7.7 have been obtained for quite extreme values of the ratio σ_h/σ_v, either very big or very small. Let us give an example when this ratio is close to unity, by using the "simplified" solution. For instance, Fig.7.19 shows the creep of a tunnel excavated in rock salt at depth h=700 m with p=0 and σ_h/σ_v=0.9. As expected, in this case the closure is faster in the horizontal direction. The dash-dot line shows the shape of the tunnel after the dimensionless time interval $k_S(t-t_o)=5\times10^6$. Two dotted lines show two intermediate positions of the walls of the tunnel at dimensionless

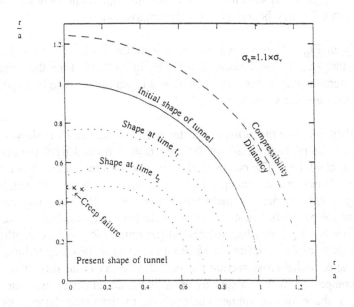

Fig.7.20 Same as in Fig.7.19 but for $\sigma_h/\sigma_v=1.1$. Closure is faster in the minor σ^P direction, where τ has high values and creep failure is also expected.

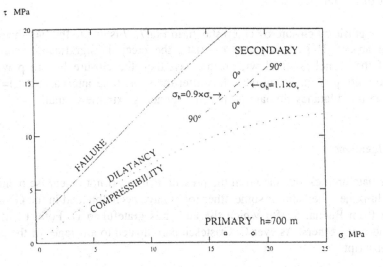

Fig.7.21 Primary and secondary stress states around a tunnel showing where τ is maximum and as such where a faster creep rate is to be expected as well as creep failure. τ is maximum in the direction of minor far field σ^P component.

times $t_1=2\times10^6$ and $t_2=4\times10^6$. Creep failure is also possible and is obviousle nonsymmetric: it occurs in the horizontal direction. The boundary of the domain up to where creep failure has spread up to time 5×10^6 is shown by ×- line. An interrupted line shows where is located the compressibility/dilatancy boundary just after excavation.

Similarly in Fig.7.20 is shown the closure of the walls of a tunnel for the case $\sigma_h / \sigma_v = 1.1$ and otherwise the same conditions as in Fig.7.19. This time the closure is faster in the vertical direction and in the same direction a creep failure is to be expected. In this figure the notations are the same as in Fig.7.19.

The explanation why the closure is faster in certain directions is shown in Fig.7.21. By a □ is marked the primary stress for the case $\sigma_h = 0.9 \ \sigma_v$ and by ◊ the corresponding primary stress for $\sigma_h = 1.1 \ \sigma_v$. The corresponding secondary stresses are shown by dash-dot lines. From here follows that the creep closure is faster there where the octahedral shear stress has a higher value, since the main terms involved in the stationary creep rate of strain component are power functions involving τ. In the same direction one has to expect creep failure. That is why depending mainly on the ratio σ_h / σ_v, on the depth and on the mechanical properties of the rock (including those involved in the creep failure) that during closure a borehole or tunnel cross section can be either longer in the direction of the major primary stress component, if the closure by creep is the dominant phenomenon compared to the change in shape of the tunnel contour due to the creep failure, or viceversa. Furthermore, if the change of the cross section shape is primarily due to creep failure, then an elargement of the cross section is to be expected in the direction in which acts the minor component of the far field stress.

The significant closure by creep shown in Fig.7.17 is due to the fact that an internal pressure is absent. If such a pressure exists, the creep is significantly curtailed. For instance, if the tunnel is filled with brine, say, then the closure by creep will slightly surpass one tenth part of the initial radius after the same time interval $k_S(t-t_o) = 7.5 \times 10^6$ as above, while the dilatancy domain around the tunnel is extremely small.

Acknowledgements

The data on rock salt shown in the present lectures, if not otherwise mentioned, are due to U. Hunsche. The data for some other rocks have been obtained in the Geomechanics Laboratory from Bucharest, for which the author is grateful to D. Fotă, E. Medveş, M. Nicolae, and many others. As ever C. Cristescu participated to all stages of the preparation of the manuscript.

REFERENCES

ABOU-SAYED, A.S., BRECHTEL, C.E. and CLIFTON, R.J. 1978. *In situ* stress determination by hydrofracturing: a fracture mechanics approach. *J.Geophys.Res.***83**,No.B6, 2851-2862.

AMADEI, B. 1983. *Rock Anisotropy and the Theory of Stress Measurements.* Springer-Verlag, Berlin.

AUBERTIN, M., GILL, D.E., LADANYI, B. 1991. A unified viscoplastic model for the inelastic flow of alkali halides. *Mech. of Mater.***11**, 63-82.

AUBERTIN, M., SGAOULA, J., GILL, D.E. 1992. A damage model for rocksalt: application to tertiary creep. *Seventh Int. Symp. on Salt.* April 6-9, 1992, Kyoto, Elsevier.

BOEHLER, J.P. 1978. Lois de comportement anisotropique des milieux continus. *Journal de Mécanique,* **17**, 153-190.

BOEHLER, J.P. and RACLIN, J. 1982. Ecrouissage anisotrope des matériaux ortotropes prédéformés. *Journal de Mécanique*, Numero special, 23-44.

BRACE, W.F. and BOMBOLAKIS, E.G. 1963. A note on brittle crack growth in compression. *J.Geophys.Res.***68**, 12, 3709-3713.

BRACE, W.F., PAULDING, B.W.Jr. and SCHOLZ, C. 1966. Dilatancy in the fracture of crystalline rocks. *J. Geophys. Res.,* **71**, 16, 3939-3953.

BUDIMAN, J.S., STURE, S. and KO, H.Y. 1992. Constitutive Behavior of Stress-Induced Anisotropic Cohesive Soil. *J. Geotechn. Engng.,* **118**, 1348.

CARTER, N.L., HANDIN, J., RUSSELL. J.E. and HORSEMAN, S.T. 1993. Rheology of rocksalt. *J. Structural Geology* (submitted).

CARTER, N.L. and HANSEN, F.D. 1983. Creep of rocksalt. *Tectonophysics,* **92**, 275-333.

CARTER, N.L., KRONENBERG, A.K., ROSS, J.V. and WILTSCHKO, D.V. 1990, Control of fluids on deformation of rocks. In: *Deformation Mechanics; Rheology and Tectonics*,Eds. R.J. Knipe and E.H. Rutter. Geol.Soc.spec.Publ., No.54, 1-13.

CHAN, K.S., BODNER, S.R., FOSSUM, A.F. and MUNSON, D.E. 1992. A constitutive model for inelastic flow and damage evolution in solids under triaxial compression. *Mechanics of Materials,* **14**, 1-14.

CHAN, K.S., BRODSKY, N.S., FOSSUM, A.F., BODNER, S.R., and MUNSON, D.E.,1993. Damage-induced nonassociated inelastic flow in rock salt. *Fourth International Symposium on Plasticity and Its Current Applications*, Baltimore, July 19-23, 1993.

CHANG, K.J., YANG, T.-W. 1982. A constitutive Model for the Mechanical Properties of Rock. *Int. J. Rock Mech. Min. Sci.*,**19**, 123-133.

CONSTANTINESCU, M. 1981. Experimental formulation of constitutive equations for solid porous materials. Ph.D. Dissertation, Bucharest.

COOLING, C.M., HUDSON, J.A., and TUNBRIDGE, L.W., 1988. *In Situ* Rock Stresses and Their Measurements in the U.K. Part II. Site Experiments and Stress Field Interpretation. *Int. J. Rock Mech. Min. Sci. & Geomech. Abstr.*, **25**, 6, 371-382.

CRISTESCU, N. 1979. A viscoplastic constitutive equation for rocks. Preprint Series in Mathematics, No.49/1979, INCREST Bucharest.

CRISTESCU, N. 1982. Rock Dilatancy in Uniaxial Tests. *Rock Mechanics*, **15**, 133-144.

CRISTESCU, N. 1985a. Fluage, Dilatance et/ou Compressibilité des Roches autour des Puits Verticaux et des Forages Pétroliers. *Revue Française de Geotechnique*, No.31, 11-22.

CRISTESCU, N. 1985b. Viscoplastic Creep of Rocks around Horizontal Tunnels. *Int. J. Rock Mech. Min. Sci.* **22**, 6, 453-459.

CRISTESCU, N. 1985c. Irreversible dilatancy or compressibility of viscoplastic rock-like materials and some applications. *Int.J.Plasticity*, **1**, 3, 189-204.

CRISTESCU, N. 1985d. Plasticity of compressible/dilatant rock-like materials. *Int.J.Engng.Sci.***23**,10, 1091-1100.

CRISTESCU, N. 1985e. Rock Plasticity. In: *Plasticity Today; Modelling, Methods and Applications*. Eds. A. Sawczuk and G. Bianchi. Elsevier Appl.Sci.Publ., Ltd., 643-655.

CRISTESCU, N. 1986. Damage and failure of viscoplastic rock-like materials. *Int. J. Plasticity*, **2**, 2, 189-204.

CRISTESCU, N. 1987. Elastic/Viscoplastic Constitutive Equations for Rock. *Int. J. Rock Mech. Min. Sci.*, **24**, 5, 271-282.

CRISTESCU, N. 1988. Viscoplastic creep of rocks around a lined tunnel. *Int.J.Plasticity*,**4**, 4, 393-412.

CRISTESCU, N. 1989a. *Rock Rheology*. Kluwer Academic Pulishers, Dordrecht, Holland, 336 pp.

CRISTESCU, N. 1989b. Plasticity of Porous Materials. In: *Proc.of Plasticity '89. The Second Int.Symp.on Plasticity and its Current Applications*. Eds. A.S.Khan and M.Tokuda, Pergamon Press, 11-14.

CRISTESCU, N. 1991a. Nonassociated elastic/viscoplastic constitutive equations for sand. *Int. J. Plasticity*, **7**, 41-64.

CRISTESCU, N. 1991b. Constitutive equations for rock salt. In: *Anisotropy and Localization of Plastic Deformation*. Proc. of Plasticity '91 Symposium, Grenoble, 12-16 Aug. 1991, Eds. J.-P. Boehler and A.S. Khan, Elsevier Appl.Sci., London, 201-204.

CRISTESCU, N. 1992. Constitutive equation for rock salt and mining applications. *Seventh Int. Symp. on Salt, April 6-9, 1992, Kyoto*, Elsevier Science Publ., Amsterdam, vol.1, 105-115.

CRISTESCU, N. 1993a. A general constitutive equation for transient and stationary creep of rock salt. *Int. J. Rock Mech.Min. Sci. & Geomech. Abstr.*, **30**,125-140.

CRISTESCU, N. 1993b. Rock Rheology. In: *Comprehensive Rock Engineering, vol.1, Rock Mechanics Principles*, Pergamon Press.

CRISTESCU, N. 1993c. Failure and creep failure around an underground opening. In: *Int. Symp. on Assessment and Prevention of Failure Phenomena in Rock Engineering*, Istanbul 5-7 April 1993, Balkema, Rotterdam , 205-210.

CRISTESCU, N. 1994. A procedure to determine nonassociated constitutive equations for geomaterials. *Int. J. Plasticity*, **10** (in press).

CRISTESCU, N. and FLOREA D. 1992. Temperature influence on the elastic/viscoplastic behaviour of bituminous concrete. *Rev. Roumaine de Mécanique Appliquée*, **37**, 603-614.

CRISTESCU, N. and HUNSCHE, U. 1992. Determination of a nonassociated constitutive equation for rock salt from experiments. In: *Finite Inelastic Deformations - Theory and Applications*. IUTAM Symp., Hannover, 19-23 Aug.1991, Eds. D. Besda and E. Stein, Springer Verlag, Berlin-Heidelberg, 511-523.

CRISTESCU, N. and HUNSCHE, U. 1993a. A constitutive equation for salt. *Proc. 7th Int. Congr. Rock Mech..*, Aachen, 16-20 Sept. 1991. Balkema, Rotterdam/Brookfield, vo.3, 1821-1830.

CRISTESCU, N. and HUNSCHE, U. 1993b. A viscoplastic model for stationary and transient creep of rock salt. *Fourth Int. Symp. on Plasticity and Its Current Applications,* Baltimore, July 19-23 (in press).

CRISTESCU, N. and HUNSCHE, U. 1993c. A comprehensive constitutive equation for rock salt: determination and application. *3rd Conf. on the Mechanical Behavior of Salt.* Paris, 1993, 177-191 (in press).

CRISTESCU, N., NICOLAE M. and TEODORESCU, S. 1989. Coal Rheology. In: *Inelastic Solids and Structures,* Pineridge Press, Swansea, 165-177.

CRISTESCU, N. and SULICIU, I. 1982. *Viscoplasticity.* Martinus Nijhoff Co, Hague - Editura Tehnica Bucuresti, 307 pp.

CUNHA, A.P. 1990. Scale effects in rock mechanics. In: *Scale Effects in Rock Masses,*Ed. A. Pinto da Cunha, Balkema, Rotterdam, 3-27.

DESAI, C.S. 1990. Modelling and testing: implementation of numerical models and their application in practice. *CISM Courses and lectures, No.311,* Springer Verlag, Wien-New York, 1-168.

DESAI, C.S., SOMASUNDARAM, S. and FRANTZISKONIS, G. 1986. A hierarhchical approach for constitutive modeling of geologic materials. *Int. J. Numer. Analyt. Meth. in Geomech.,* **10**, 3, 225-257.

DESAI, C.S., and VARADARAJAN, A. 1987. A constitutive model for quasistatic behavior of rock salt. *J. Geophys. Res.***92**, No.B11, 11445-11456.

DESAI, C.S. and ZHANG, D. 1987. Viscoplastic model for geologic materials with generalized flow rule. *Int. J. Numer. Analyt. Meth. in Geomech.,* **11**, 603-620.

DRAGON, A. and MRÓZ, Z. 1979. A Model for Plastic Creep of Rock-Like Materials Accounting for the Kinetics of Fracture. *Int. J. Rock Mech. Min. Sci. & Geomech Abstr.* **16,** 253-259.

DRUCKER, D. 1988. Conventional and Unconventional Plastic Response and Representation. *Appl. Mech. Review,* **41,** 151-162.

FISCHER, R.F., LIGHT, B.D. and PASLAY, P.R. 1992. Salt-cavern closure during and after formation. *Seventh Int. Symp. on Salt.* April 6-9, 1992, Kyoto, Elsevier Science Publ., Amsterdam.

FLOREA, DELIA. 1993a. Associated elastic/viscoplastic model for bituminous concrete. *Int. J. Engng. Sci.* (in press).

FLOREA, DELIA. 1993b. Nonassociated elastic/viscoplastic model for bituminous concrete. *Int. J. Engng. Sci.* (in press).

FRAYNE, M.A., ROTHENBURG, L. and DUSSEAULT, M.B. 1993. Four case studies in saltrock - Determination of material parameters for numerical modelling. *Proc. 3rd Conf. on Mechanical Behavior of Salt.* Paris 1993. 457-468 (in press).

GATES, D.J. 1988a. A Microscopic Model for Stress-Strain Relations in Rock - Part I. Equilibrium Equations. *Int. J. Rock Mech. Min. Sci. & Geomech. Abstr.* **25**, 6, 393-401.

GATES, D.J. 1988b. A Microscopic Model for Stress-Strain Relations in Rock - Part II. Triaxial Compressive Stress. *Int. J. Rock Mech. Min. Sci. & Geomech. Abstr.* **25**,6, 403-410.

HANSEN, F.D. and CARTER, N.L. 1980. Creep of rock salt at elevated temperature. *21st U.S. Symp. on Rock Mechanics*, May 28-30, 1980, University of Missouri - Rolla. Preprint.

HANSEN, F.D. and MELLEGARD, K.D. 1979. Quasi-static strength and deformational characteristics of domal salt from Avery Island. Topical Report RSI-0098.

HANSEN, F.D. and MELLEGARD, K.D. 1980. Creep of 50-mm Diameter Specimens of Some Salt from Avery Island, Louisiana, RE/SPEC ONWI-104.

HANSEN, F.D., MELLEGARD, K.D. and SENSENY, P.E. 1984. Elasticity and strength of ten natural rock salt. In: *The Mechanical Behavior of Salt*. Proc. First Conf. Edited by H.R. Hardy, Jr., and M. Langer. Trans Tech Publ., Clausthal-Zellerfeld, 71-83.

HERGET, G. 1987. Stress Assumptions for Underground Excavations in the Canadian Schield. *Int. J. Rock Mech. Min. Sci. & Geomech. Abstr.* **24**, 1, 95-97.

HERRMANN, W., LAUSON, H.S. 1981. Analysis of Creep Data for Various Natural Rock Salts. Sandia Rep. SAND81-2567.

HERRMANN, W., WAWERSIK, W.R. and LAUSON, H.S. 1980. Creep curves and fitting parametrs for southeastern New Mexico bedded salt. Sandia Rep. SAND-80-0087.

HETTLER, A., GUDEHUS, G. and VARDOULAKIS, I. 1984. Stress-Strain Behaviour of Sand in Triaxial Tests. In: *Results of the Int. Workshop on Constitutive Relations for Soils,* Eds. G. Gudehus, F. Darve and I. Vardoulakis, 6-8 Sept. 1982,

Grenoble. Balkema, Rotterdam, 55-66.

HORII, H. and NEMAT-NASSER, S. 1985. Compression-induced microcracks growth in brittle solids: axial splitting and shear failure. *J. Geophys. Res.* **90**, 3105-3125.

HORII, H. and NEMAT-NASSER, S. 1986. Brittle failure in compression: splitting, faulting and brittle-ductile transition. *Phil. Trans. Royal Soc. London,* **319**,1549, 337-374.

HUNSCHE, U. 1988 - Private communication

HUNSCHE, U. 1991a. Volume change and energy dissipation in rock salt during triaxial failure tests. . *Mechanics of Creep Brittle Materials 2,* Eds. A.C.F. Cocks and A.R.S. Ponter, Coll. 2-4 Sept. 1991 Leicester. Elsevier, London, 172-182.

HUNSCHE, U. 1991a - Private communication.

HUNSCHE, U. 1992a. True triaxial failure tests on cubic rock salt samples - experimental methods and results. In: *Finite Inelastic Deformations - Theory and Applications.* Proc IUTAM Symp. Hannover, Aug. 1991. Eds. D. Besdo and E. Stein. Springer Verlag, Berlin-Heidelberg, 525-536.

HUNSCHE, U. 1992b. Failure Behaviour of Rock Salt around Underground Cavities. *Sevents Int. Symp. on Salt.* April 6-9, 1992, Kyoto, Elsevier, Amsterdam , 59-65.

HUNSCHE, U. and CRISTESCU, N. 1994 (in preparation).

HUNSCHE, U., MINGERZAHN, G. and SCHULZE, O. 1993. The influence of textural parameters and mineralogical composition on the creep behavior of rock salt. *Proc. 3rd Conf. on the Mechanical Behavior of Salt,* Paris 1993 , 129-138 (in press).

HUNSCHE, U, and SCHULZE, O. 1993. Effect of humidity and confining pressure on creep of rock salt. *3rd Conf. on the Mechanical Behavior of Salt.* Paris, 223-234 (in press).

KRANZ, R.L., BISH, D.L. and BLACIC, J. D. 1989. Hydration and dehydration of zeolitic tuff from Yucca Mountain, Nevada. *Geophys. Res. Let.,* **16**, 10, 1113-1116.

LADE, P.V. and KIM, M.K. 1988. Single Hardening Constitutive Model for Frictional Materials. II. Yield Criterion and Plastic Work Contours. *Computers and Geotechnique,* **6**,13-29.

LADE, P.V., NELSON, R.B. and ITO, Y.M. 1987. Nonassociated flow and stability of granular materials. *J. Engng. Mech.,* **113**,1302-1318.

LADE, P.V. and PRADEL, D. 1990. Instability and Plastic Flow of Soil. I. Experimental Observations. *J. Engng. Mech.,* **116,** 2532.

LAJTAI, E.Z. and SCOTT DUNCAN, E.J. 1988. The Mechanism of deformation and fracture in potash rock. *J. Can. Geotechn.* **25,** 262-278.

LAJTAI, E.Z., SCOTT DUNCAN, E.J. and CARTER, B.J. 1991. The Effect of Strain Rate on Rock Strength. *Rock Mechanics and Rock Engineering,* **24,** 99-109.

LAJTAI, E.Z., SCHMIDTKE, R.H. and BIELUS, L.P. 1987. The Effect of Water on the Time-dependent Deformation and Fracture of a Granite. *Int. J. Rock Mech. Min. Sci. & Geomech. Abstr.***24,** 4, 247-255.

LAMA, R.D. and VUTUKURI, V.S. 1978. *Handbook on Mechanical Properties of Rocks.* **II** and **III.** Trans Tech Publ., Clausthal-Zellerfeld.

LINDNER, E.N., and HALPERN, J.A. 1978. *In-situ* stress in North America. A compilation. *Int.J. Rock Mech.Min.Sci.& Geomech Abstr.* **15,** 183-203.

LUX, K.H. and HEUSERMANN, S. 1983. Creep Tests on Rock Salt with Changing Load as a Basis for the Verification of Theoretical Material Laws. *Sixth International Symposium on Salt.,* 1983, vol.1, Salt Institute, 417-435.

MARTINETTI, S. and RIBACCHI, R. 1980. *In Situ* Stress Measurements in Italy. *Rock Mechanics,* **9,** 31-47.

MELLEGARD, K.D., SENSENY, P.E. and HANSEN, F.D. 1981. Quasi-Static Strength and Creep Characteristics of 100-mm diameter specimens of salt from Avery Island, Louisiana, RE/SPEC, ONWI-250.

MUNSON, D.E., DE VRIES, K.L., SCHIERMEISTER, D.M., DE YONGE W.F. and JONES R.L. 1992. Measured and calculated closures of open and brine filled shafts and deep vertical boreholes in salt. In: *Rock Mechanics.* Proc. of the 33rd U.S. Symp., Balkema, Rotterdam/Brookfield, 439-448.

MUNSON, D.E, and WAWERSIK, W.R. 1993. Constitutive modeling of salt behavior State of the technology. *Proc. Seventh Int. Congr. on Rock Mechanics,* Aachen, Sept. 1991, Balkema, Rotterdam-Brookfield, 1797-1810.

NOVA, R. 1980. The failure of transversely isotropic rocks in triaxial compression. *Int. J. Rock Mech. Min. Sci. & Geomech. Abstr.* **17,** 325-332.

REED, M.B. and CASSIE, J. 1988. Non-associated flow rules in computational plasticity.In: *Numerical Methods in Geomechanics*. Ed. G. Swoboda, Innsbruck, Balkema, Rotterdam, 481-488.

RUMMEL, F. and BAUMGARTNER, J. 1985. Hydraulic Fracturing In-Situ Stress and Permeability Measurements in the Research Borehole Kronzen, Hohen Venn (West Germany). *N. Jb. Geol. Palaont. Abh.* **171**, 183-193.

SANO, O., ITÔ, I. and TERADA, M. 1981. Influence of strain rate on dilatancy and strength of Oshima granite under uniaxial compression. *J. Geophys. Res.* **86,** No.B10, 9299-9311.

SAXENA, P.C., MOKHASHI, S.L. and RAME GOWDA, B.M. 1979. Rock stress measurements at Nagjhari tunnels, Kalinadi Hydro-electric project, India. *Fourth Int. Congr. on Rock Mechanics,* Montreux, Balkema, Rotterdam, 589-594.

SCHULZE, O. 1993. Effect of humidity on creep of rock salt. *Proc. Seventh Int. Congr. on Rock Mechanics,* Aachen, Sept. 1991, Balkema, Rotterdam-Brookfield.

SENSENY, P.E. 1985. Determination of a Constitutive Law for Salt at Elevated Temperature and Pressure. *Measurements of Rock Properties at Elevated Pressures and Temperatures, ASTM STP 869.* Eds. H.J. Pincus and E.R. Hoskins, Amer. Soc. for Testing and Materials, Philadelphia, 55-71.

SENSENY, P.E. 1986. Triaxial Compression Creep Tests on Salt From the Waste Isolation Pilot Plant. SANDIA Rep. SAND85-7261.

SENSENY, P.E., HANSEN, F.D., RUSSELL, J.E., CARTER, N.L. and HANDIN, J.W. 1992. Mechanical Behaviour of Rock salt: Phenomenology and Micromechanisms. *Int. J. Rock Mech. Min. Sci.* **29**, 363-378.

SENSENY, P.E., BRODSKY, N.S. and DeVRIES, K.L. 1993. Parameter evaluation for a unified constitutive model. *J. Engr. Mat. Technol.* ASME (submitted).

SMITH, M.B. and CHEATHAM, J.B.Jr. 1980. An anisotropic compacting yield condition applied to porous limestone. *Int. J. Rock Mech. Min. Sci. & Geomech. Abstr.* **17**, 159-165.

VAN SAMBEEK, L.L. 1992. Testing and modeling of backfill used in salt and potash mines. In: *Rock Support in Mining and Underground Construction.* Eds. P.K. Kaiser and D.R. McCreath. Balkema, Rotterdam, 583-589.

VAN SAMBEEK, L.L., FOSSUM, A., CALLAHAN, G. and RATIGAN, J. 1992. Salt Mechanics: Empirical and Theoretical Developments. *Seventh Int. Symp. on Salt,*

Kyoto, April 1992, Elsevier Science Publ., Amsterdam

VUTUKURI, V.S., LAMA, R.D. and SALUJA, S.S. 1974. *Handbook on Mechanical Properties of Rocks.* vol.1, Trans Tech Publ., Clausthal-Zellerfeld.

WANTEN, P.H., SPIERS, C.J. and PEACH, C.J. 1993. Deformation of NaCl single crystals at 0.27 T_m<T<0.44 T_m. *Proc. 3rd Conf. on Mechanical Behavior of Salt,* Paris, 1993,103-114 (in press).

WAWERSIK, W.R. and HANNUM, D.W. 1980. Mechanical behavior of New Mexico rock salt in triaxial compression up to 200°C. *J. Geophys. Res.***85,** No.B2, 891-900.

WAWERSIK, W.R. and ZEUCH, D.H. 1986. Modeling and mechanistic interpretation of creep of rock salt below 200°C. *Tectonophysics.* **121,** 125-152.

FINITE ELEMENT ANALYSIS OF TIME DEPENDENT EFFECTS IN TUNNELS

G. Gioda and A. Cividini
Polytechnic of Milan, Milan, Italy

ABSTRACT

A discussion is presented of the numerical analysis, based on the finite element method, of the time dependent effects that develop when a tunnel is driven in a rock mass characterized by a viscous behaviour. First, the so called swelling and squeezing phenomena are described considering in particular rocks containing clay minerals. Subsequently the discussion is focused on the squeezing behaviour, i.e. on the time dependent increase of the shear deformation which develops with minor volume changes. Some simple linear and nonlinear constitutive laws are presented able to describe this phenomenon and their use in the solution of boundary value problems by means of the finite element method is discussed. Finally, the results of some numerical analyses are presented that illustrate the effects taking place around tunnels driven into squeezing rocks.

1. INTRODUCTION

Various rocks and soils may show under appropriate conditions a tendency to deform during time in the presence of a constant stress state. The observed time dependent behaviour can have different mechanical causes. In fact, in some cases it depends on the multi-phase nature of the material, like in the case of primary consolidation. Or it may be the effect of the actual viscous characteristics of the skeleton, as in the case of salt rock or for the secondary compression of clay. In other instances it may be caused by the slow propagation of microcracks and faults within the material, as in the case of the "delayed" failure of hard, brittle rocks.

Here the discussion is focused on the numerical analysis of the time dependent, or viscous, behaviour of rocks, with particular reference to its influence on the deformation and stability of tunnels.

With this respect, it can be observed that time dependent effects in tunnels are often related to the presence of clay minerals in the surrounding rock [1]. There are also other types of rocks that exhibit marked time dependent effects (like e.g. the salt rock [2,3,4]) which, however, are perhaps more commonly met in mining engineering than in the construction of tunnels in the field of civil engineering.

From the engineering view point two main types of time dependent rock behavior are distinguished, that are usually referred to as "squeezing" (or creep) and "swelling" [5]. In a squeezing rock the time dependent deformation is produced, basically at constant volume, by the concentration of stresses in the vicinity of the excavation. In the case of swelling, the stress variation and the increase in water content produce a volume increment during time, which is frequently associated also to an increase in shear strains.

Note that while squeezing (or creep) can be studied even considering the rock as a one phase (solid) material, the analysis of swelling should take into account the two phase nature of the medium.

Another important distinction concerns the time required for the viscous deformations to develop. In general terms, the effects of the viscous behavior of the medium surrounding a tunnel can be subdivided into short term and long term effects [6].

Short term effects occur at a rate comparable to the rate of tunnel advancing. They develop mainly in the vicinity of the excavation face and in many cases become appreciable before the installation of the permanent support. Short term effects seem predominantly associated to squeezing and in a numerical analysis they should be studied by actually simulating the progress of excavation [6,7]. If simplified two-dimensional calculations are performed, the analysis should consider the longitudinal section of the tunnel along its axis in axisymmetric conditions.

Long term effects show up after a relatively long time from the completion of excavation. Probably they are more common in swelling rocks but, depending on the rate of excavation and on the characteristics of the rock, they can be also present in squeezing rocks. Usually long term effects are more or less uniformly distributed along a considerable portion of the tunnel. When a simplified, two-dimensional analysis of these effects is attempted, it is preferable to assume a plane strain condition normal to the tunnel axis.

In the following the discussion is subdivided into three main parts. First, squeezing and swelling phenomena are described considering in particular rocks containing clay minerals. Subsequently the discussion is limited to the squeezing behaviour and some simple linear and nonlinear constitutive laws are presented able to describe it. The use of these laws in the solution of boundary value problems through the finite element method is also discussed. Finally, some finite element calculations are presented that illustrate the effects that develop around tunnels driven in creeping rock.

2. SQUEEZING (OR CREEP) AND SWELLING BEHAVIOR

The squeezing behaviour can be defined [8,9] as the increase of the shear deformation of a rock element during time, when the element is subjected to a constant deviatoric stress state. This phenomenon can be also seen as a slow development of plastic strains and usually produces a limited volumetric deformation. Consequently, the volume and the water con-

tent of the element remain practically constant during time. This phenomenon is generally encountered in rocks having a high percentage of
micaceous particles or containing clay minerals with low swelling
capacity, such as Illite and Kaolinite.

It should be noted that often the development of time dependent deformation at constant volume is also referred to as creep.

When a tunnel is driven into a soft squeezing rock, the ground advances slowly into the opening, without visible fracturing or loss of
continuity. Soft clays show this behaviour also in shallow tunnels, while
stiff clays and rocks show it at higher depths. In this later case, however, squeezing may show up in combination with raveling. In fact, the
continuos inward deformation of the rock may take place together with the
separation of rock fragments, or blocks, from the roof and walls of the
excavation.

At a microscopic level the causes of squeezing are similar to those
at the base of the secondary compression of clays. They can be found in
the redistribution of the contact forces between the particles induced by
the shear stresses. This, in turn, produces reorientation, relative movements of the particles, and perhaps bending and breakage of them and, as
a macroscopic effect, a slow time dependent movement of the rock toward
the opening.

Swelling can be defined as the time dependent volume increase of a
rock element caused by the stress release and/or by an increase of its
water content. In a clayey rock the absorption of water is associated
with an increase of the distance between the solid particles that, in
turn, produces a reduction of the interaction forces connecting them.

The decrease of the particle bonds reduces the overall shear
resistance of the rock and permits the development of shear deformation.
As a consequence, swelling may develop together with time dependent deviatoric strains, or squeezing.

Note that the causes influencing the interaction forces between the
particles [10], such as the thickness of the double layer, the type of
cations in the pore water, the unbalanced electric charge of the particles, etc. can influence also the development of swelling and creep
phenomena.

Since swelling is related to a variation of the water content, the duration of this phenomenon and its rate are markedly influenced by the permeability of the clay.

Swelling characterizes highly overconsolidated clays with plasticity index greater than 30, or clayey rocks, that contain clay minerals such as Montmorillonite, which has a high swelling capacity.

Squeezing soils have a consistency from soft to medium and their natural water content is rather high, close to the liquid limit. On the other hand swelling clays are overconsolidated and at least moderately stiff. Their consistency is from medium to high and their natural water content is rather low, close to the plastic limit or less.

Note that the tendency to a volume increase through absorption of water can be found also in rocks which do not contain clay minerals, such as Anhydrite. In this case swelling is caused by the hydration process that transforms Anhydrite into Gypsum, which has a larger volume per unit mass than Anhydrite.

The swelling potential can be quantitatively assessed by means of one dimensional (oedometer) tests. A sample of swelling clay is first consolidated under a pressure q_0, representing the in situ vertical stress at rest (fig.1). At the end of consolidation (or when primary consolidation is completed and some part of secondary compression is developed), the water content has reached a value of w_0. Then, the load is rapidly reduced to zero, and the possibility is given to the sample to absorb water. Swelling initiates and an increase of the water content during the time is measured. If no loads are applied to the sample, swelling continues until the water content reaches a final value w_2.

If during swelling, at time t_1, further the vertical displacement is constrained (in a tunnel this condition may correspond to the installation of the permanent liner), the water content remains constant at a value w_1 and the vertical pressure approaches an asymptotic value q_1.

Various laboratory procedures have been suggested in the literature in order to estimate the load exerted by swelling rocks on tunnel liners. Among them, the popular procedure for performing swelling tests [11] is summarized here. The sample is first subjected in the oedometer to a load-unload-reload cycle in a dry state, i.e. without allowing it to absorb water, in order to eliminate the disturbance caused by sampling and test setting up operations. This stage of the tests is represented by curves a,b,c in fig.2 and leads to a state represented by point D'. In fig.2, σ_D is the in situ vertical stress condition and ϵ_B is assumed to represent the vertical strain that would be possible to measure in situ in case of complete unloading without any swelling.

At the end of this stage, the sample is allowed to swell by absorbing water and the vertical stress is kept constant and equal to σ_D. Due to swelling the vertical strain increases up to point D.

When swelling under constant vertical load terminates, the load is reduced by subsequent steps, thus obtaining the so called "swelling" curve s. It can be observed that the swelling under constant vertical load (from D' to D in fig.2) is not constant since curves c and s tend to meet at point A. If reloading is carried on until the vertical stress corresponding to point A is reached, it can be expected that the amount of swelling under constant vertical load tends to zero.

These results can be used either for estimating the deformation of the rock caused by the excavation of the tunnel, or for evaluating the load acting on the liner of a tunnel driven in a swelling rock.

The squeezing, or creep, characteristics of a rock are usually evaluated through unconfined compression tests or by triaxial tests [12]. In tunneling practice the determination of the squeezing potential of a rock is sometime based on the so-called "stability number" N_t which is the ratio between the in situ pressure at the tunnel depth and the unconfined compression strength of the rock. It has been observed, in fact, that an empirical correlation exists between this number and the overall behaviour of a tunnel excavated in squeezing rock:

$N_t < 1$ The soil practically behaves as an elastic material. No problems due to squeezing are expected.

$1 < N_t < 4$ Some squeezing is observed, but the rate of movement is very slow. No major problems related to squeezing are anticipated.

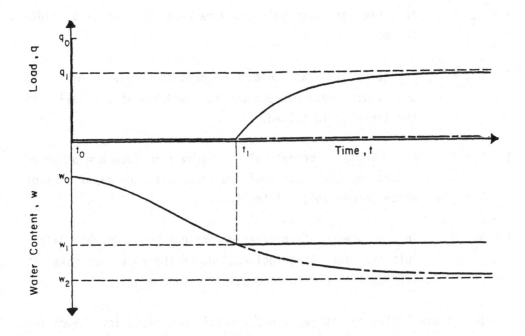

Fig.1. Variation of vertical stress and water content vs. time for swelling soils.

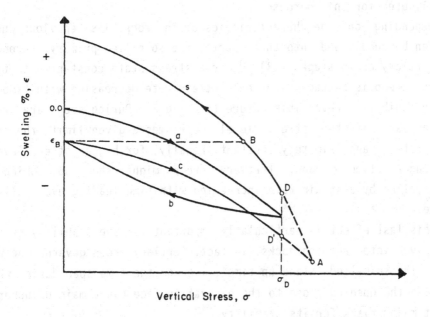

Fig.2. Vertical stress vs. strain in Huder-Amberg's one dimensional swelling test.

$4 < N_t < 5$ The use of a tunnel boring machine (TBM) or of a shield becomes necessary.

$5 < N_t < 6$ Squeezing is fast enough to produce problems in filling the anular void between the rock and the shield tail when the liner is installed.

$6 < N_t < 7$ Appreciable movements of the excavation face are observed caused by deformation of the rock mass, the major part of which occurs ahead of the face.

$7 < N_t$ The progress of excavation presents considerable difficulties. The large deformation of the rock mass makes it difficult to control the shield advancement.

The above limits do not represent a sufficient basis for the design of tunnels in squeezing rock and in most cases an adequate experimental investigation is needed. As previously observed, long term triaxial creep tests, in which the samples are allowed to deform during time, are customarily used for this purpose.

Depending on the characteristics of the rock, its behaviour during time can be subdivided into three parts, the so called primary, secondary and tertiary creep stages [13]. Under a stress state constant with time, primary creep is characterized by a strain rate decreasing with time and usually exhibits a reversible nature (see fig.3). During secondary creep, that appears if the stress level overcomes a given limit, the creep strain rate is approximately constant. Finally, for higher stress levels, or longer time spans, tertiary creep might show up, which is characterized by a strain rate increasing with time leading eventually to failure.

This last effect is particularly important for the stability of tunnels driven into creeping rocks. In fact, tertiary creep governs the value of the so called "stand up time", i.e. of the time span during which part of the opening close to the excavation face can remain unsupported without major risks for its stability.

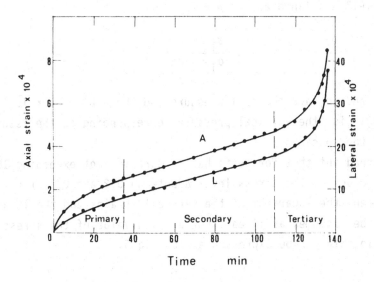

Fig.3. Axial (A) and lateral (L) strain vs. time curves from constant
load compression tests.

In the following Sections some creep laws will be discussed which
have been used for the analysis of opening in creeping rock.

3. AN EMPIRICAL CREEP LAW

Various laws, based on the interpretation of laboratory creep tests,
have been presented in the literature for interpreting the time dependent
behaviour of rocks [14-16]. Here the law proposed by Semple, Hendron and
Mesri [16] will be discussed which is based on the results of a series of
consolidated, undrained triaxial creep tests performed on remolded specimens of clay gouge. For some tests the vertical load was mantained constant during time, while in other tests the load was increased by subsequent steps.

The recorded vertical strain ϵ_1 versus time t data were plotted
using both standard and logarithmic scales (figs.4,5,6). The curves corresponding to the various tests where characterized by the value of the

"stress level" Δ defined as

$$\Delta = \frac{\sigma_1 - \sigma_3}{\sigma_{1\ell} - \sigma_3} \quad . \tag{1}$$

Here, σ_3 and σ_1 are the cell pressure and the applied vertical stress, and $\sigma_{1\ell}(\sigma_3)$ is the vertical pressure corresponding to the istantaneous failure of the specimen.

It turned out that for tests having duration not exceeding 100 hours and for values of the stress level Δ between 0.2 and 0.8, the relationship between the logarithm of the vertical strains and the logarithm of time may be assumed as linear. For a given value of the stress level Δ this relationship can be expressed as (cf.fig.6)

$$\epsilon_1 = \bar{\epsilon}_1 \, (t/\bar{t})^{\lambda} \quad , \tag{2}$$

where λ is a material constant, $\bar{t}$ is a chosen reference time and $\bar{\epsilon}_1$ is the vertical deformation measured at the reference time.

Within the mentioned range of Δ, and for a chosen value of $\bar{t}$ (say 1 hour) an almost linear relationship exists between the logarithm of $\bar{\epsilon}_1$ and the stress level Δ (cf.fig.7)

$$\log\left[\bar{\epsilon}_1(\Delta)\right] = \log\left[\bar{\epsilon}_1(\Delta=0)\right] + \beta\Delta \quad ,$$

which is equivalent to

$$\bar{\epsilon}_1 = Be^{\beta\Delta} \quad . \tag{3}$$

By substituting eq.(3) into eq.(2), the following creep law is reached

$$\epsilon_1(t) = Be^{\beta\Delta} \, (t/\bar{t}_1)^{\lambda} \quad , \tag{4}$$

where β, λ and B are material parameters. Since eq.(4) has been developed considering the stress level Δ constant with time, this law cannot be

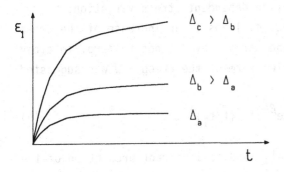

Fig.4. Qualitative representation of creep strain vs. time plots from constant load compression tests.

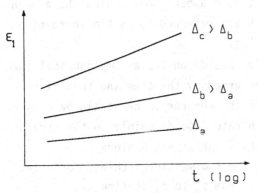

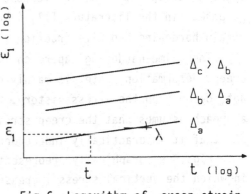

Fig.5. Creep strain vs. logarithm of time plots for diagrams in fig.4.

Fig.6. Logarithm of creep strain vs logarithm of time plots for diagrams in fig.4.

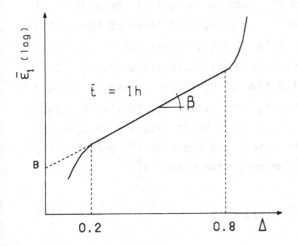

Fig.7. Variation of the creep strain with the stress level Δ for time equal to 1 hour.

directly adopted in the case of a time dependent stress variation.

The authors observed that eq.(4) leads to an increase of the creep strains during time even when the stress level tends to zero. To avoid this erroneous prediction, a modified form of the creep law was suggested

$$\epsilon_1(t) = B \ (e^{\beta\Delta}-1) \ (t/\bar{t}_1)^{\lambda} \ .$$

(5)

Note that eq.(5) could still lead to incorrect prediction of the viscous rock behaviour for high stress levels, exceeding about 80%.

As previously observed, empirical creep laws like eq.(5) cannot be directly applied when stresses change during time as a function of the unknown creep strains. To overcome this drawback some "rules" have been proposed in the literature [17] which are referred to as time-hardening, strain-hardening and life fraction rules.

The time-hardening approach is based on the assumption that the creep deformation rate is mainly governed by the time and that it does not depend on the stress history. On the contrary, the strain-hardening approach assumes that the creep strain rate depends mainly on the strains and that it is practically independent of the stress history.

Figs.8,9 graphically represent the two mentioned rules, for a test in which the vertical stress increases from σ_1^a to σ_1^b at time t_1.

The third approach, the so called life fraction rule, represents a compromise between the two previous methods and can be outlined as follows. It is assumed that constant load creep tests with different values of the vertical stress, say σ_1^a and σ_1^b, lead to failure of the samples at time t_{fa} and t_{fb}, respectively. Consider a creep test in which the vertical load is kept constant and equal to σ_1^a until a vertical strain ϵ_1^a is reached at time t^a. Then the vertical stress is increased up to σ_1^b. It is assumed that the ratio between the time at which the stress is changed and the time needed to reach failure t_f, or total life, remain constant. As a consequence the following relationship holds (fig.10) between the time t^a, at which the stress is increased, and the time t^b, defining the corresponding point on the strain-time curve for stress σ_1^b

$$t^b = t^a \ t_{fb}/t_{fa} \ .$$

(6)

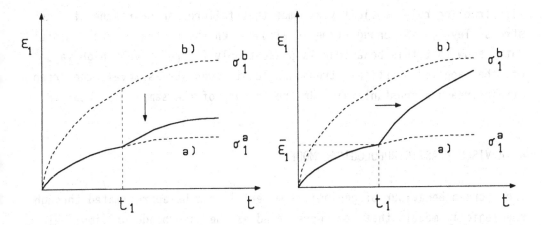

Fig.8. Schematic representation of the time-hardening rule.

Fig.9. Schematic representation of the strain-hardening rule.

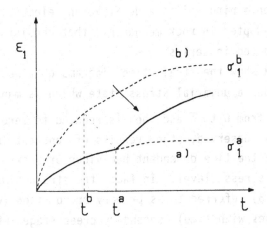

Fig.10. Schematic representation of the life fraction rule.

The practical use of this last rule is perhaps more complex than that of the previous ones. In fact, the experimental evaluation of the total life of the sample could involve non negligible problems, especially if the applied vertical stress is much smaller than the unconfined compression strength of the material. This is due to the fact that the

life fraction rule implicitly assumes that failure can be reached for any
stress level, if enough time is allowed. On the contrary, experimental
data show that this behaviour is present only for relatively high values
of the applied vertical stress, while at lower stress levels the creep
strains reach a constant value and no failure of the sample is observed.

4. A VISCO-PLASTIC RHEOLOGICAL MODEL

The creep behaviour of geological materials can be approximated through
rheological models that are represented as the assemblage of simple elements, such as springs (Hooke elements), dashpots (Newton elements) and
frictional-cohesive (de St.Venant) elements.

Some of the simplest rheological models are depicted in fig.11:
Kelvin and Maxwell models, consisting of springs and dashpots, and
Bingham model, containing also a de St.Venant element. Sometimes also
Burgers model is adopted in rock mechanics, that consists of Maxwell and
Kelvin models connected in series.

Figs.12 and 13 show the strain-time diagrams obtained by Maxwell and
Burgers models under a uniaxial stress state which is mantained constant
in a time interval from 0 to $\bar{t}$ and then is reduced to zero.

As previously observed, the results of constant load compression
tests indicate that the time dependent behaviour of rocks is markedly influenced by the stress level. In fact, the strain-time diagrams show
three zones (fig.3) referred to as primary creep stage (where the creep
strain rate decreases with time), secondary creep·stage (where the strain
rate is approximately constant with time) and tertiary creep stage (in
which the strain rate increases with time leading to failure). Only transient (primary) creep is present at low stress levels. For higher stress
levels also secondary and tertiary creeps show up. Under purely volumetric stress the time dependent behaviour is mainly transient, or primary.

The above observations indicate that a rheological model for rocks
should account for the three creep stages previously mentioned and that,
even in the simple isotropic case, the model should separately consider
the volumetric and the deviatoric behaviour.

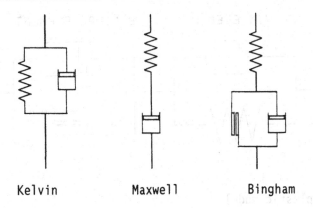

Kelvin   Maxwell   Bingham

Fig.11. Some rheological models used in geomechanics.

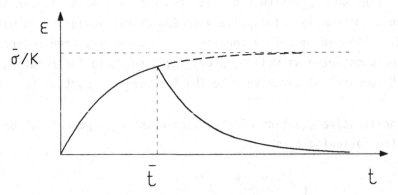

Fig.12. Kelvin model: creep strain vs. time curve under uniaxial stress $\bar{\sigma}$

kept constant from 0 to $\bar{t}$ and then reduced to zero.

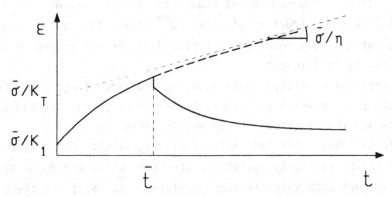

Fig.13. Burgers model, other data as in fig. 12.

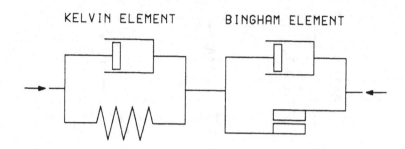

Fig.14. Visco-plastic model.

These requirements are met by the model depicted in fig.14. It derives from the one proposed in [18] and consists of a visco-elastic Kelvin model, accounting for primary creep, connected in series with a visco-plastic Bingham model accounting for secondary creep. Tertiary creep can be considered as well by providing a suitable law relating the values of the mechanical parameters to the irreversible part of the creep strains.

The constitutive equation of the visco-elastic component can be expressed in the following form

$$\underline{\sigma}_t = \underline{D}^{ve} \, \underline{\varepsilon}_t^{ve} + \underline{V}^{ve} \, \underline{\dot{\varepsilon}}_t^{ve} \quad , \tag{7}$$

where $\underline{\sigma}_t$ is the stress vector, $\underline{D}^{ve}$ and $\underline{V}^{ve}$ are the elastic and viscous constitutive matrices, $\underline{\varepsilon}^{ve}$ is the recoverable creep strain vector and t denotes the time. A superposed dot means time derivative. Note that matrix $\underline{V}^{ve}$ has the same structure of matrix $\underline{D}^{ve}$ where the elastic parameters, e.g. shear and bulk moduli, are replaced by the corresponding shear and bulk viscosity coefficients.

As to the visco-plastic behaviour, let consider first for sake of simplicity the case in which the material parameters of the Bingham model are constants and its frictional component is perfectly plastic.

A basic hypothesis adopted in the following is that the Newton element (dashpot) carries only deviatoric stresses, while the de St.Venant element can support both volumetric and deviatoric stresses. On the basis

of this assumption a stress state exceeding the resistance of the frictional-cohesive element can be subdivided into its octahedral component
σ_{oct}, that acts on the frictional element, and its deviatoric part $\underline{s}$
which, in turn, is expressed as the sum of the components carried by the
dashpot $\underline{s}^{cr}$ and by the de St.Venant element $\underline{s}^y$

$$\underline{\sigma}_t = \underline{m} \cdot (\sigma_{oct})_t + \underline{s}_t^y + \underline{s}_t^{cr} \ . \tag{8}$$

In the above equation $\underline{m}$ is a vector the entries of which are equal
to 1, if they correspond to normal stresses, or to 0, if they correspond
to shear stresses. Note that $\underline{s}^y$ is readily evaluated on the basis of the
octahedral stress since the resulting stress $\underline{\sigma}^y$ must fulfill the yield
condition F adopted for the de St.Venant element

$$\underline{\sigma}_t^y = \underline{m} \cdot (\sigma_{oct})_t + \underline{s}_t^y \ . \tag{9}$$

The visco-plastic (creep) strain rate $\underline{\dot{\epsilon}}^{cr}$ is subdivided into its volumetric $\dot{\epsilon}_{vol}^{cr}$ and deviatoric $\underline{\dot{e}}^{cr}$ parts, which are represented by orthogonal vectors

$$\underline{\dot{\epsilon}}_t^{cr} = \underline{m} \cdot (\dot{\epsilon}_{vol}^{cr})_t/3 + \underline{\dot{e}}_t^{cr} \ , \tag{10}$$

$$\underline{m}^T \ \underline{\dot{e}}_t^{cr} = 0 \ . \tag{11}$$

In eq.(11) a superposed T means transpose. Since the viscous and co-
hesive-frictional elements (cf.fig.14) are subjected to the same deformation, plastic and creep strains coincide. Consequently the plastic flow
rule is expressed in terms of the creep strain rate $\underline{\dot{\epsilon}}^{cr}$ and of the rate
of the plastic multiplier λ through the standard relationship

$$\underline{\dot{\epsilon}}_t^{cr} = \lambda_t \ \underline{n}_t \ , \tag{12}$$

where $\underline{n}$ is the gradient of the plastic potential Q determined at $\underline{\sigma}^y$

$$\underline{n}_t = \frac{\partial Q}{\partial \underline{\sigma}_t^y} \ . \tag{13}$$

By substituting eq.(12) into eq.(10), and taking into account eq.
(11), the following expression of the rate of the plastic multiplier is
readily obtained

$$\lambda_t = \frac{(\underline{\dot{e}}_t^{cr})^T \, \underline{\dot{e}}_t^{cr}}{(\underline{\dot{e}}_t^{cr})^T \, \underline{n}_t} \quad , \tag{14}$$

where vector $\underline{n}$ is given by eq.(13) and the deviatoric creep strain rate
$\underline{\dot{e}}^{cr}$ depends through the constitutive relationship of the dashpot on the
deviatoric stresses $\underline{s}^{cr}$

$$\underline{s}_t^{cr} = \underline{V}^{cr} \, \underline{\dot{e}}_t^{cr} \quad . \tag{15}$$

Note that due to the assumption of purely deviatoric behaviour of
the dashpot, the viscosity matrix $\underline{V}^{cr}$ depends on only one deviatoric viscosity coefficient.

The governing equation for non linear, time dependent problems
discretized into finite elements can be written in the following general
form, where the summation runs over the elements of the mesh, $\underline{B}$ is the
strain-nodal displacement matrix and $\underline{u}$ is the nodal displacement vector,

$$\Sigma \int_V \underline{B}^T \, \underline{\sigma}(\underline{u}_t) \, dV - \underline{f}_t =$$

$$= \underline{a}(\underline{u}_t) - \underline{f}_t = \underline{0} \quad . \tag{16}$$

Eq.(16) represents the equilibrium condition between the total
applied forces, collected in vector $\underline{f}$, and those equivalent in the finite
element sense to the total stresses $\underline{\sigma}$.

In principle the non linear problem (16) could be solved, through an
algorithm for the minimization of non linear functions, by searching for
the displacement vector $\underline{u}$ that once introduced in eq.(16) brings to zero
the difference between vectors $\underline{a}$ and $\underline{f}$. In practice, however, non linear
problems are solved through one of two main techniques, the so called
initial stress and initial strain methods [19], among which the most convenient procedure in the presence of time dependent behaviour turns out

to be the initial strain technique. This implies that at any time t an extra term is present in the load vector which depends on the current creep strains through the following relationship

$$\underline{\sigma}_t = \underline{D} \ [\underline{\varepsilon}_t - \underline{\varepsilon}^V(\underline{\sigma}_t)] \ , \tag{17}$$

where $\underline{D}$ is the elastic constitutive matrix governing the instantaneous response of the material and $\underline{\varepsilon}^V$ is the total (visco-elastic and visco-plastic) creep strain.

A straightforward iterative scheme for time integration can be based on the assumption of constant strain rate during a time increment Δt, leading to the following approximation of the viscous strain increment for the i-th iteration

$$(\Delta \underline{\varepsilon}_t^V)^i = (\underline{\varepsilon}_t^V)^i - \underline{\varepsilon}_{t-\Delta t}^V =$$

$$= \Delta t \ [\vartheta \ (\underline{\dot{\varepsilon}}_t^V)^{i-1} + (1-\vartheta) \ \underline{\dot{\varepsilon}}_{t-\Delta t}^V] \ . \tag{18}$$

In eq.(18) ϑ is a coefficient the value of which can vary between 0 and 1. If $\vartheta=0$ the creep strains depend only on the relevant quantities at the end of the preceding time increment, thus leading to an explicit integration scheme.

Here an implicit Crank-Nicholson scheme has been adopted, with $\vartheta=1/2$, which requires the following main operations to be carried out in each iteration:

a) Displacements, strains and stresses are evaluated at the beginning of the i-th iteration through a linear elastic analysis

$$\underline{K} \ \Delta \underline{u}_t^i = \underline{r}_t^{i-1} \ ,$$

$$\underline{u}_t^i = \underline{u}_t^{i-1} + \Delta \underline{u}_t^i \quad , \quad \underline{\varepsilon}_t^i = \underline{\varepsilon}_t^{i-1} + \Delta \underline{\varepsilon}_t^i \ ,$$

$$\Delta \underline{\sigma}_t^i = \underline{D} \ [\Delta \underline{\varepsilon}_t^i - (\Delta \underline{\varepsilon}_t^V)^i] \quad , \quad \underline{\sigma}_t^i = \underline{\sigma}_t^{i-1} + \Delta \underline{\sigma}_t^i \ .$$

In the above equations $\underline{D}$ is the elastic constitutive matrix and $\underline{K}$ is the corresponding stiffness matrix. For the first iteration (i=1) it is assumed that

$$(\dot{\underline{\varepsilon}}_t^V)^0 = \dot{\underline{\varepsilon}}_{t-\Delta t}^V \quad ,$$

regardless the value of ϑ, and the vector of residual forces $\underline{r}$ is

$$\underline{r}_t^0 = \Delta\underline{f}_t + \int_V \underline{B}^T \underline{D}_t \ [(\dot{\underline{\varepsilon}}_t^V)^0 - \dot{\underline{\varepsilon}}_{t-\Delta t}^V)] \ dV =$$

$$= \Delta\underline{f}_t + \int_V \underline{B}^T \underline{D}_t \ (\Delta t \ \dot{\underline{\varepsilon}}_{t-\Delta t}^V) \ dV \quad ,$$

where $\Delta\underline{f}_t$ is the increment of external loads for the current time increment.

b)   The viscous strains $(\underline{\varepsilon}_t^V)^i$ at time t

$$(\underline{\varepsilon}_t^V)^i = (\underline{\varepsilon}_t^{ve})^i + (\underline{\varepsilon}_t^{cr})^i \quad ,$$

are evaluated by means of the visco-elastic and visco-plastic relationships, and of eq.(18) where ϑ is equal to 1/2,

$$(\underline{\varepsilon}_t^{ve})^i = [\underline{D}^{ve} + \tfrac{2}{\Delta t} \underline{V}^{ve}]^{-1} \cdot \left([(\underline{\sigma}_t)^i - \underline{\sigma}_{t-\Delta t}] + \right.$$

$$\left. - [\underline{D}^{ve} - \tfrac{2}{\Delta t} \underline{V}^{ve}] \cdot \underline{\varepsilon}_{t-\Delta t} \right)$$

and

$$(\underline{\varepsilon}_t^{cr})^i = \underline{\varepsilon}_{t-\Delta t}^{cr} + \tfrac{1}{2} [\lambda_t^i - \lambda_{t-\Delta t}] \cdot [\underline{n}_t^i + \underline{n}_{t-\Delta t}]$$

where

$$\lambda_t^i = \lambda_{t-\Delta t} +$$

$$+ 2 \cdot \frac{[(\underline{e}_t^{cr})^i - \underline{e}_{t-\Delta t}^{cr}]^T \cdot [(\underline{e}_t^{cr})^i - \underline{e}_{t-\Delta t}^{cr}]}{[(\underline{e}_t^{cr})^i - \underline{e}_{t-\Delta t}^{cr}]^T \cdot [\underline{n}_t^i + \underline{n}_{t-\Delta t}]}$$

and

$$\underline{n}_t^i = \frac{\partial Q}{\partial (\underline{\sigma}_t^y)^i}$$

$$(\underline{e}_t^{cr})^i = \underline{e}_{t-\Delta t}^{cr} + \frac{\Delta t}{2} (\underline{V}^{cr})^{-1} \cdot [(\underline{s}_t^{cr})^i + \underline{s}_{t-\Delta t}^{cr}]$$

c)   The unbalanced nodal forces are computed

$$\underline{r}_t^i = \Sigma \int_V \underline{B}^T \underline{D} [(\underline{\varepsilon}_t^v)^i - (\underline{\varepsilon}_t^v)^{i-1}] \, dV \quad .$$

If a convergence test is fulfilled (e.g. based on the norm of the unbalanced force vector) the analysis continues with the next time increment, otherwise a further iteration is performed.

As previously observed, the possible development of tertiary creep can be accounted for by providing a suitable law governing the variation of the visco-plastic parameters with increasing strains. To this purpose the simple relationship graphically depicted in fig.15 can be adopted. The shear strength (c, ϕ) and viscosity (η) parameters of Bingham model are considered as functions of a scalar measure of the irreversible deformation represented by the square root of the second invariant J_2^{cr} of the deviatoric creep strains $\underline{e}^{cr}$. The material parameters keep their peak values until the deviatoric deformation reaches a first limit value J_p. Then a linear reduction of the parameters takes place with increasing plastic strains, leading to their residual values when the deviatoric deformation reaches a second limit J_r.
      The iterative procedure previously outlined can still be adopted if the material parameters are assumed to be constant within each time step, and their values are evaluated on the basis of the average irreversible deformation during the step. In order to avoid the possible instability of the integration process caused by the reduction of the viscosity coefficient, the time step length is automatically reduced when an increase of the residual forces is observed during the iterations.

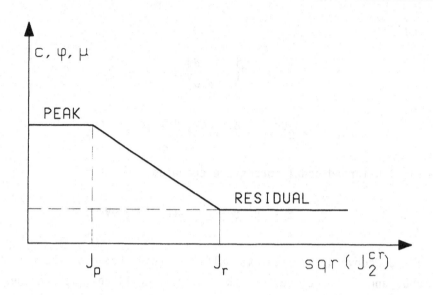

Fig.15. Variation of the parameters of the visco-plastic model with
deviatoric plastic strains.

     In order to check the accuracy of the numerical technique, an ana-
lytical solution has been worked out for a plane strain compression test
on a visco-plastic material element obeying a perfectly plastic Tresca
yield condition [20]. The sample is subjected to known vertical and hori-
zontal pressures σ_1, σ_2, and to an out of plane strain ϵ_3, such that the
yield condition can be expressed in the following form

$$\sigma_1^y - \sigma_2^y + \sigma_0 = 0 \quad ,$$

where σ_0 is the yield limit. Under this assumption it is possible to
show that only linear relationships, with constant coefficients, exist
between the applied stresses, the out of plane stress σ_3 and the part of
the stress state fulfilling the yield criterion. This permits to solve
analytically the visco-plastic problem through the well known correspondence principle based on Laplace transform. The details of the derivation
are not presented here for sake of briefness.
     A comparison between analytical (solid lines) and numerical (dots)
results is shown in fig.16.

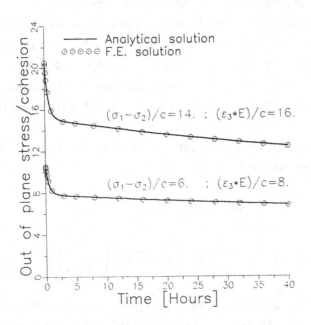

Fig.16. Comparison between analytical and numerical simulation of plane strain compression test.

5. SOME NUMERICAL EXAMPLES

It has been already observed that the time dependent deformation strongly influences the overall behaviour of tunnels driven into ,squeezing rocks [21] and that numerical analyses are particularly suitable to investigate this phenomenon [7,22,23]. In the following the results of a series of finite element calculations are presented which aim to show some of the effects of viscosity on tunnels.

In a first group of analyses [6] the rock is assumed to behave according to the visco-elastic Kelvin model. Usually simple rheological models do not give an acceptable approximation of the squeezing behaviour, especially for long time periods or when a significant change of the stress level is expected. However these models can give a reasonable results if short time effects are studied, and if a limited variation of the stress level during time is expected.

In the calculations the time dependent shear deformation are governed by Kelvin model, while the volumetric behaviour of the isotropic rock is linearly elastic. Fig.17 shows a comparison between the shear strain-time data from triaxial creep test and the corresponding curves obtained with the adopted model, with different values of the material parameters.

The shear strain-time relationships of this rheological model, for a shear stress τ constant with time, are

$$\gamma(t) = \gamma_e + \gamma_v(t) \quad , \tag{19}$$

$$\gamma_e = \tau/G_1 \quad , \quad \gamma_v(t) = \frac{\tau}{G_2} \left[1-\exp\left(-\frac{G_2}{\eta} t\right)\right] \quad . \tag{20a,b}$$

In the above equations the suffix e and v denote, respectively, elastic and viscous strains; G_1 is the istantaneous shear modulus and G_2 is the shear modulus of Kelvin model and η is its viscosity.

Two sets of finite element analyses have been performed. In both cases axisymmetric conditions about the tunnel axis were assumed and the processes of excavation and liner installation were simulated. In the first set of "fully lined" analyses no gap between the liner and the face of the excavation exists. In the second set of "partially lined" analyses the distance between the tunnel face and the leading edge of the liner is equal to one tunnel radius.

Various factors influence the amount of creep deformation around the tunnel: the tunnel diameter D; the rate af advancing R_a, expressed in diameters per unit time, and the mechanical properties, shear modulus G and viscosity η, of the rock. Through these quantities the parameter λ is defined

$$\lambda = \frac{G_T D}{\eta R_a (1-\alpha)} \quad . \tag{21}$$

In eq.(21) G_T represents the "long term" shear modulus of the rock, that depends on G_1 and G_2,

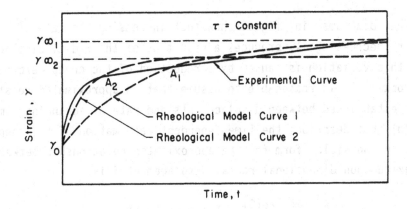

Fig.17. Visco-elastic   approximation   of   the   creep   strain-time
relationship.

$$G_T = (1/G_1 + 1/G_2)^{-1}$$

and α is the ratio between G_T and G_1

$$\alpha = G_T/G_1 \quad .$$

The parameter λ varies between 0 and ∞. $\lambda \to 0$ corresponds to a very
fast excavation, the rate of which tends to infinity $R_a \to \infty$, when no time
is allowed for creep deformation during the tunneling process. In this
case only instantaneous strains and displacements are present, governed
by the shear modulus G_1. $\lambda \to \infty$ indicates a very slow rate of advancing,
tending to zero, i.e. $R_a \to 0$. The entire creep deformation caused by a step
of excavation develops before the next step initiates and at any time the
displacements depend only on the shear modulus G_T.

Fig.18 shows the variation along the tunnel axis z of the radial
displacement u_r of the tunnel perimeter, for both fully lined and partially lined cases. The displacements have been made non dimensional
multiplying them by a factor $M_c/\gamma Ha$, where M_c is the confined modulus of
the rock depending on its istantaneous shear G_1 and bulk B_1 moduli, γ is
the rock own weight per unit volume, H and a are the depth of tunnel and
its radius.

The diagrams in fig.18 show that increasing the rate of advancing R_a, or decreasing λ, produces a reduction of the radial displacements. Since this variation is caused by a reduction of the creep deformation in the rock, it is reasonable to assume that an approximated relationship can be established between displacements and rate of advancing similar to eq.(20b) that describes the time dependent deformation of the rheological model. A possible form for the approximated relationship between λ and the maximum non dimensional radial displacement δ is

$$\frac{\delta(\lambda)-\delta(0)}{\delta(\infty)-\delta(0)} = 1 - \exp(-C_\delta \cdot \lambda) \ . \tag{22}$$

The non dimensional constant C_δ can be evaluated introducing the values of δ obtained by the finite element analyses into eq.(22) rewritten as follows

$$C_\delta = -\frac{1}{\lambda} \ln \left[1 - \frac{\delta(\lambda)-\delta(0)}{\delta(\infty)-\delta(0)} \right] \ . \tag{23}$$

The values of the maximum non dimensional radial displacements and of the corresponding coefficients C_δ are reported in Table I.

Table I. Maximum radial displacement δ and coefficients C_δ

λ	Fully lined δ	C_δ	Partially lined δ	C_δ
0.00	0.32	-	1.99	-
1.09	0.47	0.61	2.45	0.28
2.18	0.54	0.57	2.75	0.26
3.27	0.58	0.56	2.97	0.25
∞	0.63	-	3.75	-

Fig.19 shows the comparison between the finite element results, in terms of maximum non dimensional displacement versus λ, and the corresponding quantity obtained by eq.(22) in which the average values of C_δ for fully lined (0.58) and partially lined (0.26) cases were introduced.

A reasonable agreement exists between the two sets of data. This indicates that equations similar to eq.(22) could be adopted for a rapid estimation of the squeezing effects around a tunnel, under the assumption of viscoelastic behaviour for the rock.

Another problem that should be considered when driving a tunnel in a squeezing rock is the increase of radial displacements when the continuous process of excavation is arrested. To designate the time elapsed after the arrest of excavation the non dimensional time ξ is adopted

$$\xi = t \frac{G_2}{\eta} . \qquad (24)$$

Figs.20 and 21 shows the variation with time ξ of the radial displacements for the fully lined and partially lined cases. It should be pointed out that these results do not consider the development of irreversible strains in the rock mass and the possible loss of strength due to the increase of shear strains. Consequently they could underestimate the actual deformation of the squeezing rock.

An attempt to consider the mentioned effects [24] was made adopting the visco-plastic model described in Section 4 in the analysis of a deep circular tunnel. The time dependent behaviour of the rock is illustrated in fig.22 by the vertical creep strain vs. time diagrams obtained through the numerical simulation of triaxial creep tests under constant vertical load. In this figure, τ_F is the maximum shear stress carried by the material without undergoing secondary creep effects; $\epsilon_{1\ell}$ is the maximum vertical strain reached when $\tau=\tau_F$; $t_\ell(90\%)$ is the time at which 90 per cent of $\epsilon_{1\ell}$ develops when $\tau=\tau_F$.

The tunnel problem is treated in plane strain and axisymmetric conditions, a cylindrical reference system (r,ϑ,z) is adopted, where z denotes the direction normal to the deformation plane. The only non vanishing displacement component is the radial one $u(r)$ and the stress and total

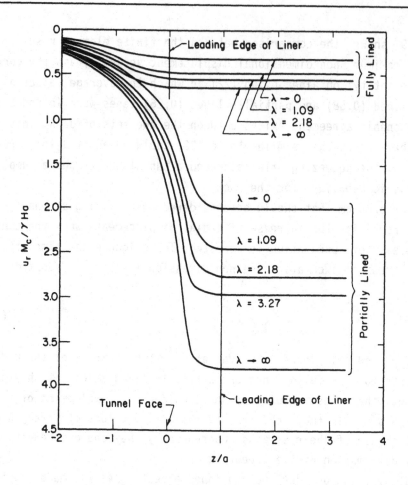

Fig.18. Radial displacement for fully and partially lined tunnels.

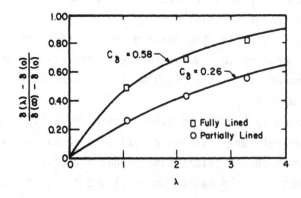

Fig.19. Relationship between λ and maximum radial displacement, δ.

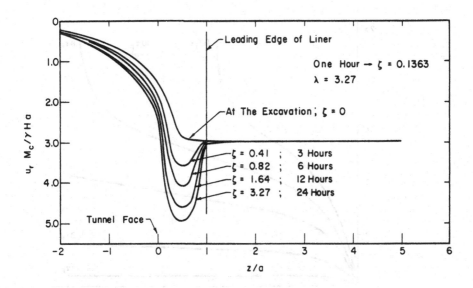

Fig.20. Radial displacement for partially lined tunnel after a temporary interruption of excavation.

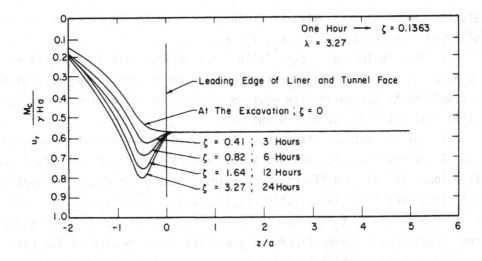

Fig.21. Radial displacement for fully lined tunnel after a temporary interruption of excavation.

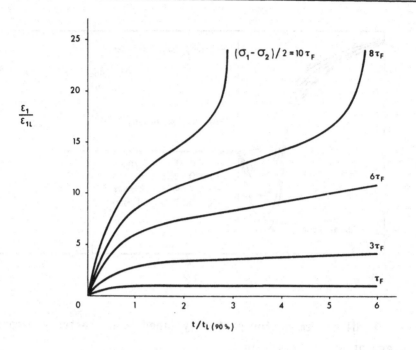

Fig.22. Creep strain vs. time from the numerical simulation of constant
load triaxial creep tests.

(elastic and viscous) strain fields are completely described by the
following components: σ_r, σ_ϑ, σ_z, ϵ_r, ϵ_ϑ.

At the beginning of calculations the stress state in the rock mass
surrounding the tunnel is constant and isotropic. The tunnel excavation
is simulated by decreasing the uniform pressure on its perimeter from the
initial value to zero at a constant rate.

Various situations have been considered at the end of excavation.
Case I refers to the unlined situation. Cases II to V refer to lined
situations in which different time spans separate the end of excavation
(t_e) from the liner installation (t_i): case II t_i=1.5t_e; III t_i=1.0t_e;
IV t_i=0.5t_e and V t_i=t_e. Two sub-cases, denoted by subscripts a and b,
were considered for each lined analyses. The liner considered for case b
has a stiffness which is twice as large as the stiffness of the liner in
case a.

The results of these calculations are shown in fig.23, in terms of
radial displacement vs. time, and in fig.24, in terms of plastic radius

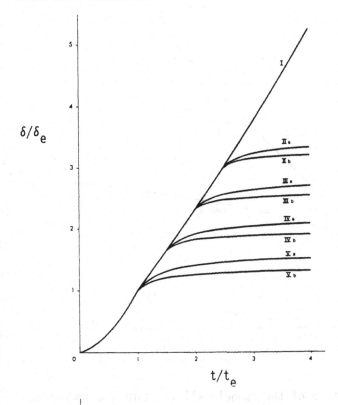

Fig.23.
Inward displacement δ of the tunnel wall vs. time, for the unlined case I and for the lined cases II to V.

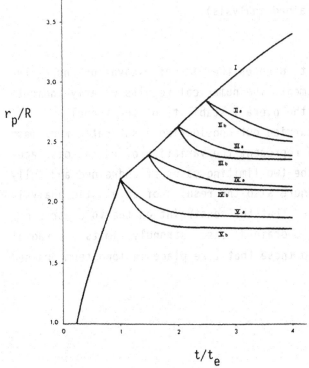

Fig.24.
Plastic radius r_p vs. time, for the unlined case I and for the lined cases II to V.

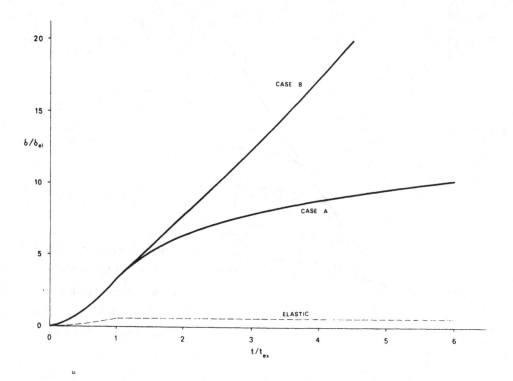

Fig.25. Inward displacement δ of the tunnel wall vs. time (A, undrained
analysis; B, fully drained analysis).

vs. time. In these figures, t_e denotes the time of excavation and δ_e the
corresponding radial displacement. The numerical results clearly indicate
the effect of the support on the overall stability of the tunnel.

Similar analyses were carried out considering a saturated rock mass
[25]. The diagrams in fig.25 represent the variation of radial displacement vs. time obtained for the two limiting cases of undrained and fully
drained behaviour. In this figure also the results of an elastic analysis
are shown. It can be observed that the constraint on the volumetric deformation existing in the undrained case strongly limits the radial
displacements with respect to those that take place in long-term drained
conditions.

6. CONCLUSIONS

Some aspects have been discussed of the finite element analysis of tunnels driven into squeezing rocks, i.e. rocks that exhibit a viscous behaviour associated with the deviatoric stress state.

After describing the major characteristics of the phenomenon, some creep laws proposed in the literature have been presented. Among them, those based on linear and non linear rheological models are particularly suitable for analyses carried out through the finite element method.

The results of a series of finite element calculations show that these laws permit to consider the effects of the so called primary, secondary and tertiary creep stages in a stress analysis, taking also into account the possible loss of strength of the material with increasing deviatoric deformation.

This type of calculations could be useful for the engineer during the design of tunnels or underground openings and also for the "back analysis" of in situ measurements performed during the excavation, in order to reach a better understanding of the overall time dependent behaviour of the rock mass.

ACKNOWLEDGEMENTS

The financial support of the Italian Ministry of the University and Research and the help of Mrs. Paola Nava are gratefully acknowledged.

REFERENCES

[1] Aiyer A.K., An analytical study of the time-dependent behaviour of underground openings, Ph.D.Thesis, University of Illinois at Urbana-Champaign, Department of Civil Enginering, 1969.

[2] Fossum A.F., Visco-plastic behaviour during the excavation phase of a salt cavity, Int.J.Numer.Anal.Methods in Geomechanics, Vol.1, 45-55, 1977.

[3] Cristescu N., Rock rheology, Kluwer Academic Publ., Dordrecht, 1989.

[4] Ottosen N.S., Viscoelastic-viscoplastic formulas for analysis of cavities in rock salt. Int.J.Rock Mech.Min.Sci.& Geomech.Abstract, Vol.23, 201-212, 1986.

[5] Nakano R., On the design of water tunnels in relation with the type and magnitude of rock load with special references to the mechanism and prediction of squeezing-swelling rock pressure, Bulletin of the National Research Institute of Agricultural Engineering, Ministry of Agriculture and Forestry (Japan), No.12, 89-142, 1974.

[6] Ghaboussi J. and G. Gioda, On the time dependent effects in advancing tunnels, Int. J.Numer.Anal.Methods in Geomech., Vol.1, 1977.

[7] Sakurai S., Approximate time-dependent analysis of tunnel support structure considering progress of tunnel face. Int.J.Num.Anal.Meth. Geomech., Vol.2, 159-175, 1978.

[8] Terzaghi K., Soft ground tunneling, in From Theory to Practice in Soil Mechanics, J. Wiley & Sons, 338-357, 1960.

[9] Peck R.B., Deep excavation and tunneling in soft ground,, Proc.7th ICSMFE, Mexico City, State-of-the-art volume, 225-284, 1969.

[10] Mitchell J.K., Fundamentals of soil behavior, J.Wiley & Sons, 1976.

[11] Einstein H.H. and N. Bishoff, Design of tunnels in swelling rock, Proc.16th Symp.on Rock Mechanics, Minneapolis, 120-130, 1975.

[12] Bishop A.W. and H.T. Lowenburg, Creep characteristics of two undisturbed clays, Proc.7th ICSMFE, Vol.1, Mexico City, 1969.

[13] Cividini A., Constitutive behaviour and numerical modeling, in Comprehensive Rock Engineering (Hudson J.A. et al. edts.), Pergamon Press, Oxford, Vol.1, 395-426, 1993.

[14] Christiansen R.W. and T.H. Wu, Analysis of clay deformation as a rate process, J.Soil Mech.Found.Div., ASCE, Vol.90, No.SM6, 125-157, 1964.

[15] Singh A. and J.K. Mitchell, General stress-strain-time function for soils, J.Soil Mech.Found.Div., ASCE Vol.94, No.SM1, 21-46, 1968.

[16] Semple R.M., A.J. Hendron and G. Mesri, The effects of time-dependent properties of altered rock on tunnel support requirements, U.S. Department of Transportation, Federal Railroad Administration, Rep. No. FRA-OR&D 74-30, December, 1973.

[17] Nair K. and A.P. Boresi, Stress analysis for time-dependent problems in rock mechanics, Proc.2nd ICRM, Beograd,, Vol.2 531-536, 1970.

[18] Gioda G., A finite element solution of non-linear creep problems in rocks, Int.J.Rock Mech.Min.Sci.& Geomech.Abst., Vol.18, 35-46, 1981.

[19] Zienkiewicz O.C., I.C. Cormeau, Viscoplasticity-plasticity and creep in elastic solids, Int.J.Numer.Meth.Engng., Vol.8, 821-845, 1974.

[20] Cividini A., G. Gioda and A. Carini, A finite element analysis of the time dependent behaviour of underground openings, Proc.7th Int. Conf.on Computer Methods and Advances in Geomech., Cairns, 1991.

[21] Landanyi B., Time dependent response of rock around tunnels, in Comprehensive Rock Engineering (Hudson J.A. et al. edts.), Pergamon Press, Oxford, Vol.2, 77-112, 1993.

[22] Christian J.T. and B.J. Watt, Undrained visco-elastic analysis of soil deformation, Proc.Symp.on Application of Finite Elem.Method in Geotechnical Engineering, Waterways Experiment Station, Vicksburg, Missisipi, 1972.

[23] Cristescu N., Viscoplastic creep of rocks around horizontal tunnels, Int.J.Rock Mech.Min.Sci.& Geomech.Abstr, Vol.22, No.6, 453-459, 1985

[24] Gioda G., On the non linear squeezing effects around circular tunnels, Int.J.Numer.Anal.Methods in Geomech., Vol.6, 1982.

[25] Gioda G. and A. Cividini, Viscous behaviour around an underground opening in a two-phase medium, Int.J.Rock Mech.Min.Sci.& Geomech. Abstract, Vol.18, 1981.

COUPLED ANALYSES IN GEOMECHANICS

D.V. Griffiths
Colorado School of Mines, Golden, CO, USA

ABSTRACT

This paper presents a thorough review of the implementation of the Biot equations of equilibrium for a saturated elastic porous medium into a finite element code. The generation of the matrix equations via a Galerkin formulation is described in detail and an incremental form of the equations is presented suitable for nonlinear analysis. Simple elasto-plastic models for geomaterials are described including a survey of different failure criteria and the Viscoplastic algorithm for stress redistribution is reviewed. Full listings of computer code in modular form are included for both elastic and elasto-plastic examples and attention is drawn to certain programming features which lead to improved efficiency and verstility. The paper concludes with several examples of finite element analysis of transient collapse problems of relevance to geotechnical engineering.

Chapter 1

Biot Formulation

The term 'coupled' has tended to be associated with analyses in which the dependent variables are not all of the same type. In the case of the transient behaviour of saturated soils or rocks the variables in question become 'displacements' and 'excess pore pressures'. This chapter develops the Biot equilibrium and continuity equations in one- and two-dimensions, and later chapters will describe how coupled soil behaviour can be incorporated into a finite element code enabling transient analysis of boundary value problems.

1.1   One–dimension

The consolidation of soil in one–dimension is the simplest example of a 'viscous' soil model as it can be represented by the rheological idealisation shown in figure 1.1.

Under oedometer conditions with a maximum drainage path length of D, the effective stress σ' and the excess pore pressure u_w are shown in figure 1.2 at a typical time t.

Both σ' and u_w are functions of y and t and at all times the effective stress equation:

$$\sigma = \sigma' + u_w \tag{1.1}$$

is satisfied. From figure 1.2 it is clear that at any time t, the equilibrium of an element of soil is given by:

$$\frac{\partial \sigma'}{\partial y} + \frac{\partial u_w}{\partial y} = 0 \tag{1.2}$$

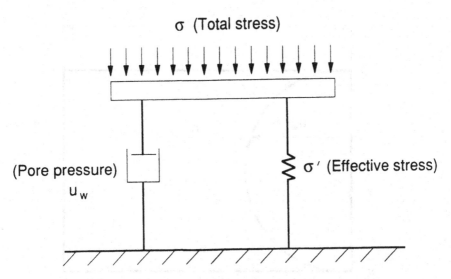

Figure 1.1: **Spring–dashpot analogy for 1–d consolidation**

Introducing the *coefficient of volume compressibility* where

$$m = \frac{(1+\nu')(1-2\nu')}{E'(1-\nu')}$$

we get:

$$\frac{\partial \sigma'}{\partial y} = \frac{1}{m}\frac{\partial \epsilon}{\partial y} = \frac{1}{m}\frac{\partial^2 v}{\partial y^2} \qquad (1.3)$$

hence:

$$\frac{1}{m}\frac{\partial^2 v}{\partial y^2} + \frac{\partial u_w}{\partial y} = 0 \qquad (1.4)$$

where v is the vertical displacement of the soil element and $\epsilon = \partial v/\partial y$ is the axial (small) strain.

The net flow rate into (or out of) an element of soil must equal the rate of change of volume of that element, hence:

$$\frac{\partial}{\partial t}\frac{\partial v}{\partial y} + \frac{k}{\gamma_w}\frac{\partial^2 u_w}{\partial y^2} = 0 \qquad (1.5)$$

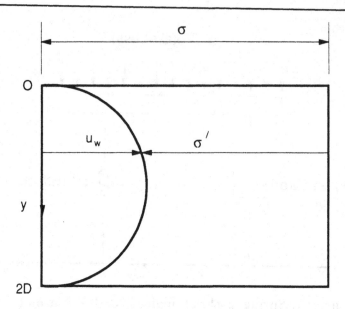

Figure 1.2: **Isochrone of pore pressure under oedometer conditions**

where k is the permeability and γ_w is the unit weight of water. It may be noted that the right term of eqn. 1.5 comes from the Laplacian operator and the left term from the first time derivative of axial strain. Equations 1.4 and 1.5 represent the *coupled* equations for a one–dimensional poro–elastic medium in terms of independent variables (y, t) and dependent variables (v, u_w).

One–dimensional consolidation is special however, in that the total stress remains constant throughout. This means that the equations can be rearranged in such a way that the displacement term v is eliminated. First we note from eqn. 1.1 that since σ is constant, then at any depth y the rate of change of effective stress must be equal and opposite to the rate of change of excess pore pressure, thus:

$$\frac{\partial \sigma'}{\partial t} + \frac{\partial u_w}{\partial t} = 0 \tag{1.6}$$

Multiplying eqn. 1.6 by m we get:

$$m\frac{\partial \sigma'}{\partial t} = -m\frac{\partial u_w}{\partial t} \tag{1.7}$$

but from the stress/strain relationship:

$$m\frac{\partial \sigma'}{\partial t} = \frac{\partial \epsilon}{\partial t} = \frac{\partial}{\partial t}\frac{\partial v}{\partial y} \qquad (1.8)$$

hence from 1.7 and 1.8, eqn. 1.5 becomes:

$$\frac{k}{m\gamma_w}\frac{\partial^2 u_w}{\partial y^2} = \frac{\partial u_w}{\partial t} \qquad (1.9)$$

Finally, defining the *coefficient of consolidation* $c = k/(m\gamma_w)$ we get the classical Terzaghi one–dimensional consolidation equation:

$$c\frac{\partial^2 u_w}{\partial y^2} = \frac{\partial u_w}{\partial t} \qquad (1.10)$$

Equation 1.10 is an *uncoupled* partial differential equation in which the excess pore pressure u_w is the only dependent variable. The one–dimensional case is generally solved in this form because the displacement, strain and effective stress at any depth and at any time can be easily retrieved once the excess pore pressure is known.

1.2   Two–dimensions

A coupled formulation has more practical importance when we consider two–dimensional problems since it now becomes necessary to combine the total stress with continuity of the soil mass.

Figure 1.3 shows the variation in (total) direct stress and shear stresses across a 'small' element of dimensions $(\delta x, \delta y)$.

From equilibrium considerations, we get:

$$\frac{\partial \sigma_x}{\partial x} + \frac{\partial \tau_{xy}}{\partial y} + X = 0$$
$$\frac{\partial \tau_{xy}}{\partial x} + \frac{\partial \sigma_y}{\partial y} + Y = 0 \qquad (1.11)$$

where X and Y are 'body forces' and complementary shears τ_{xy} and τ_{yx} are assumed to be equal.

From the principal of effective stress (eqn. 1.1) in both the $x-$ and $y-$ directions:

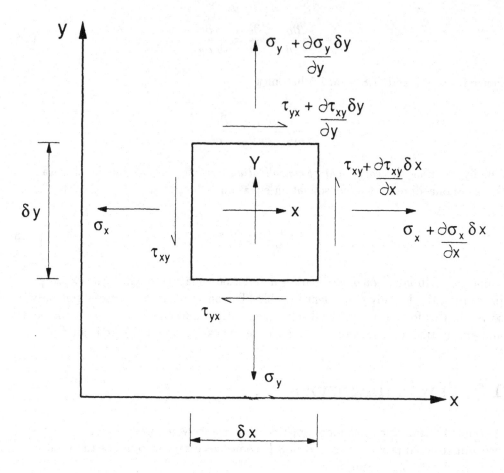

Figure 1.3: **Equilibrium of a 2–d element of soil** ⨠

$$\sigma_x = \sigma_x' + u_w$$

$$\sigma_y = \sigma_y' + u_{u'}$$

(1.12)

eqn. 1.11 can be written as:

$$\frac{\partial \sigma_x'}{\partial x} + \frac{\partial \tau_{xy}}{\partial y} + \frac{\partial u_w}{\partial x} = 0$$

$$\frac{\partial \tau_{xy}}{\partial x} + \frac{\partial \sigma_y'}{\partial y} + \frac{\partial u_w}{\partial y} = 0$$

(1.13)

Assuming plane strain conditions, small strains and ignoring body forces, the stress terms in eqn. 1.13 can be eliminated in terms of displacements to give:

$$\frac{E'(1-\nu')}{(1+\nu')(1-2\nu')}\left[\frac{\partial^2 u}{\partial x^2}+\frac{(1-2\nu')}{2(1-\nu')}\frac{\partial^2 u}{\partial y^2}+\frac{1}{2(1-\nu')}\frac{\partial^2 v}{\partial x\partial y}\right]+\frac{\partial u_w}{\partial x}=0$$

$$\frac{E'(1-\nu')}{(1+\nu')(1-2\nu')}\left[\frac{1}{2(1-\nu')}\frac{\partial^2 u}{\partial x\partial y}+\frac{\partial^2 v}{\partial y^2}+\frac{(1-2\nu')}{2(1-\nu')}\frac{\partial^2 v}{\partial x^2}\right]+\frac{\partial u_w}{\partial y}=0$$

$$(1.14)$$

We now consider two–dimensional continuity in that the net inflow (or outflow) must equal the change in volume of the element of soil thus:

$$\frac{\partial}{\partial t}\left(\frac{\partial u}{\partial x}+\frac{\partial v}{\partial y}\right)+\frac{k_x}{\gamma_w}\frac{\partial^2 u_w}{\partial x^2}+\frac{k_y}{\gamma_w}\frac{\partial^2 u_w}{\partial y^2}=0\qquad(1.15)$$

As was observed in the development of the one–dimensional equations, the second and third terms come from the Laplacian operator and the first term from the first time derivative of volumetric (small) strain ϵ_v, where in plane strain:

$$\epsilon_v = \epsilon_x + \epsilon_y$$

$$= \frac{\partial u}{\partial x}+\frac{\partial v}{\partial y}\qquad(1.16)$$

Equations 1.14 and 1.15 represent the *coupled* or Biot (1941) equations for a two–dimensional poro–elastic medium in terms of independent variables (x, y, t) and dependent variables (u, v, u_w). It is these equations which will form the basis of the majority of solutions presented in the remainder of this paper.

1.3   Uncoupled two–dimensional theory

Coupling of the pore pressure and displacement terms in multi–dimensional problems occurs because the total stress at a point generally changes as the soil mass strains during consolidation (e.g. Mandel–Cryer effect). If the total stress is assumed to remain constant as is done in a number of approximate two– and three-dimensional solutions, a pseudo consolidation theory emerges which leads to an uncoupled form of the equations in terms of the excess pore pressure only. This method involves the elimination of the displacement components (u, v) from the coupled equations.

Firstly we note that under plane strain conditions:

$$\sigma'_x = \frac{E'(1-\nu')}{(1+\nu')(1-2\nu')}\left\{\epsilon_x + \frac{\nu'}{1-\nu'}\,\epsilon_y\right\}$$

$$\sigma'_y = \frac{E'(1-\nu')}{(1+\nu')(1-2\nu')}\left\{\epsilon_y + \frac{\nu'}{1-\nu'}\,\epsilon_x\right\} \tag{1.17}$$

and from eqn. 1.12, if there is no change in total stress, any change in excess pore pressure u_w must be immediately counterbalanced by a change in effective stress, thus:

$$\frac{\partial\sigma'_x}{\partial t} + \frac{\partial u_w}{\partial t} = 0$$

$$\frac{\partial\sigma'_y}{\partial t} + \frac{\partial u_w}{\partial t} = 0 \tag{1.18}$$

Substituting from eqn. 1.17 we get:

$$\frac{E'(1-\nu')}{(1+\nu')(1-2\nu')}\,\frac{\partial}{\partial t}\left\{\epsilon_x + \frac{\nu'}{1-\nu'}\,\epsilon_y\right\} + \frac{\partial u_w}{\partial t} = 0$$

$$\frac{E'(1-\nu')}{(1+\nu')(1-2\nu')}\,\frac{\partial}{\partial t}\left\{\epsilon_y + \frac{\nu'}{1-\nu'}\,\epsilon_x\right\} + \frac{\partial u_w}{\partial t} = 0 \tag{1.19}$$

which after adding and rearranging gives:

$$\frac{E'}{(1+\nu')(1-2\nu')}\,\frac{\partial}{\partial t}(\epsilon_x + \epsilon_y) + 2\,\frac{\partial u_w}{\partial t} = 0 \tag{1.20}$$

Finally we express ϵ_x and ϵ_y in terms of small strains to give:

$$\frac{\partial}{\partial t}\left(\frac{\partial u}{\partial x} + \frac{\partial v}{\partial y}\right) = -\frac{2(1+\nu')(1-2\nu')}{E'}\,\frac{\partial u_w}{\partial t} \tag{1.21}$$

and then substitute this expression into eqn. 1.15 to give the *uncoupled* equation:

$$c_x\frac{\partial^2 u_w}{\partial x^2} + c_y\frac{\partial^2 u_w}{\partial y^2} = \frac{\partial u_w}{\partial t} \tag{1.22}$$

where the *coefficients of consolidation* in the $x-$ and $y-$ directions under plane strain conditions are defined:

$$c_x = \frac{k_x E'}{2\gamma_w(1+\nu')(1-2\nu')}$$

$$c_y = \frac{k_y E'}{2\gamma_w(1+\nu')(1-2\nu')}$$

(1.23)

Equation 1.22 is the two-dimensional counterpart of eqn. 1.10.

Solution of the uncoupled equations enables the transient behaviour of pore pressures to be computed given a set of initial conditions. It may be noted that this equation also governs transient temperature changes over a conductor if the material properties c_x and c_y are interpreted as thermal conductivity values.

Chapter 2

Finite element discretisation

In this section, the coupled equations (eqns. 1.14, 1.15) are discretised using the finite element method into a matrix form suitable for numerical computations of real boundary value problems.

2.1   Matrix form

The significance of the *coupled* formulation is highlighted when the order of discretisation (or finite element type) for excess pore pressures is different to that for displacements of the solid skeleton. A popular scheme (see e.g. Zienkiewicz 1977) uses 8–node quadrilaterals for the displacements and 4–node quadrilaterals for the excess pore pressures as shown in figure 2.1. This type of 'overlay' will lead to 20 degrees of freedoms per element; 16 nodal displacements held in the vector $\hat{\mathbf{r}}$ and 4 excess pore pressures held in the vector $\hat{\mathbf{u}}_{\mathbf{w}}$.

The 'hat' notation is used to denote discretised as opposed to actual values of the dependent variables. In the following development, a similar notation to that adopted by Hicks (1990) has been been employed.

The displacement field in terms of nodal values takes the form:

$$\hat{\mathbf{e}} = \mathbf{N}\hat{\mathbf{r}} \tag{2.1}$$

where:

$$\hat{\mathbf{e}} = [\, \hat{u} \; \hat{v} \,]^{T}$$

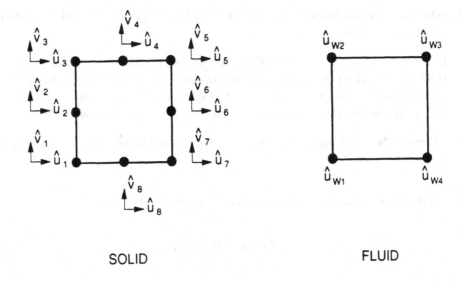

SOLID   FLUID

Figure 2.1: **8–node/4–node overlay for 2–d coupled problems**

$$\hat{\mathbf{r}} = [\; \hat{u}_1 \; \hat{v}_1 \; \hat{u}_2 \; \hat{v}_2 \; \cdots \; \hat{u}_8 \; \hat{v}_8 \;]^T$$

and:

$$\mathbf{N} = \begin{bmatrix} N_1 & 0 & N_2 & 0 & \cdots & N_8 & 0 \\ 0 & N_1 & 0 & N_2 & \cdots & 0 & N_8 \end{bmatrix}$$

Discretisation of the excess pore pressure field in terms of nodal values takes the form:

$$\hat{u}_w = \mathbf{M}\hat{\mathbf{u}}_\mathbf{w} \qquad (2.2)$$

where:

$$\hat{\mathbf{u}}_\mathbf{w} = [\; \hat{u}_{w1} \; \hat{u}_{w2} \; \hat{u}_{w3} \; \hat{u}_{w4} \;]^T$$

and:

$$\mathbf{M} = [\; M_1 \; M_2 \; M_3 \; M_4 \;]$$

We now follow a Galerkin procedure in which we:

1. substitute the discretised versions of u, v and u_w from eqn. 2.1 and 2.2 into the coupled equations 1.14 and 1.15,

2. In order to minimise the residual $\mathcal{R}$, 'weight' the equations by multiplying through by each of the shape functions in turn; note that eqn. 1.14 being the equilibrium statement is weighted by the shape functions N, and eqn. 1.15 being the continuity statement is weighted by the shape functions M,

3. integrate the resulting expressions over the area of the element and equate to zero.

The equilibrium equation 1.14 can be written in the matrix form:

$$A^T D^e A e + E u_w = f \tag{2.3}$$

in which:

$$A = \begin{bmatrix} \partial/\partial x & 0 \\ 0 & \partial/\partial y \\ \partial/\partial y & \partial/\partial x \end{bmatrix}$$

$$D^e = \frac{E'(1-\nu')}{(1+\nu')(1-2\nu')} \begin{bmatrix} 1 & \dfrac{\nu'}{(1-\nu')} & \dfrac{\nu'}{(1-\nu')} \\ \dfrac{\nu'}{(1-\nu')} & 1 & \dfrac{\nu'}{(1-\nu')} \\ 0 & 0 & \dfrac{(1-2\nu')}{2(1-\nu')} \end{bmatrix}$$

$$E = [\,\partial/\partial x \ \partial/\partial y\,]^T$$

Following substitution of the discretised terms, the following equation is obtained:

$$A^T D^e A N \hat{r} + E M \hat{u}_w - f = \mathcal{R} \tag{2.4}$$

Application of Galerkin's method to this equation using the displacement shape functions N as the weighting coefficients leads to the first matrix equation:

$$k_m \hat{r} + c \hat{u}_w = f \tag{2.5}$$

where the familiar solid stiffness matrix k_m (16 rows, 16 cols) is given by:

$$k_m = \int_{V^e} B^T D^e B \, d(\text{vol}) \qquad (2.6)$$

and B is the usual element strain/displacement matrix. Any externally applied nodal forces have also been included in equation 2.5 in the vector f.

The 'coupling' matrix c (16 rows, 4 cols) is given by:

$$c = \int_{V^e} N^T H \, d(\text{vol}) \qquad (2.7)$$

where

$$H = EM$$

(Note: c is termed the 'coupling' matrix since it contains elements of both the displacement and excess pore pressure discretisation.)

Similarly, the continuity equation 1.15 can be written in the matrix form:

$$\frac{d}{dt} E^T e + E^T K E \, u_w = 0 \qquad (2.8)$$

where the permeability tensor is given by,

$$K = \frac{1}{\gamma_w} \begin{bmatrix} k_x & 0 \\ 0 & k_y \end{bmatrix}$$

in which k_x and k_y are the principal permeabilities in the x- and y- directions.

Following substitution of the discretised terms:

$$\frac{d}{dt} E^T N \hat{r} + E^T K E M \hat{u}_w = \mathcal{R} \qquad (2.9)$$

application of Galerkin's method using the pore pressure shape functions M as the weighting coefficients leads to the second matrix equation:

$$c^T \frac{d\hat{r}}{dt} - k_p \hat{u}_w = 0 \qquad (2.10)$$

where the familiar fluid 'stiffness' or Laplacian matrix k_p (4 rows, 4 cols) is given by:

$$k_p = \int_{V^e} H^T K H \, d(\text{vol}) \tag{2.11}$$

and

$$c^T = \int_{V^e} H^T N \, d(\text{vol}) \tag{2.12}$$

2.2   Incremental equations and time discretisation

Although not essential in elastic analyses due to the principal of superposition, the matrix equations 2.5 and 2.10 are conveniently expressed in an *incremental* form which enables changes in the displacements and pore pressures to be computed in response to incremental changes in loading. The incremental form is certainly needed for nonlinear analyses, so its incorporation into the numerical algorithm leads to a more general formulation.

We also need to consider methods for advancing the equations in time, so a relatively low–order finite difference scheme of the 'θ' type is suggested.

Our starting point for the incremental approach is a reiteration of the matrix equations derived in the previous section, hence

$$k_m \hat{r} + c \, \hat{u}_w = f \tag{2.13}$$

$$c^T \frac{d\hat{r}}{dt} - k_p \hat{u}_w = 0 \tag{2.14}$$

Assume that the solution for displacements and pore pressures is known at two time intervals t^i and t^{i+1} where $\Delta t = t^{i+1} - t^i$. We denote the values of the dependent variables at these time stations by the respective superscripts i and $i+1$.

Consider eqn. 2.13 written at these two time intervals, thus:

$$k_m \hat{r}^i + c \, \hat{u}_w{}^i = f^i \tag{2.15}$$

$$k_m \hat{r}^{i+1} + c \, \hat{u}_w{}^{i+1} = f^{i+1} \tag{2.16}$$

The first incremental equation is given by subtraction as follows:

$$k_m \, \Delta \hat{r}^i + c \, \Delta \hat{u}_w^{\ i} = \Delta f^i \qquad (2.17)$$

where $\Delta \hat{r}^i = \hat{r}^{i+1} - \hat{r}^i$, $\Delta \hat{u}_w^{\ i} = \hat{u}_w^{i+1} - \hat{u}_w^i$ and $\Delta f^i = f^{i+1} - f^i$

Regarding the time derivative in eqn. 2.14, we note that for small Δt:

$$\hat{r}^{i+1} = \hat{r}^i + \Delta t \left[(1 - \theta)\frac{d\hat{r}^i}{dt} + \theta \frac{d\hat{r}^{i+1}}{dt} \right] \qquad (2.18)$$

where $0 \leq \theta \leq 1$ is a scalar parameter which weights the time derivative at the beginning and end of the time interval.

We now write eqn 2.14 at the two time intervals, thus:

$$c^T \frac{d\hat{r}^i}{dt} - k_p \, \hat{u}_w^{\ i} = 0 \qquad (2.19)$$

$$c^T \frac{d\hat{r}^{i+1}}{dt} - k_p \, \hat{u}_w^{\ i+1} = 0 \qquad (2.20)$$

By multiplying eqn. 2.19 by $\Delta t(1 - \theta)$ and eqn. 2.20 by $\Delta t \theta$ and adding, we get:

$$\Delta t \, c^T \left[(1 - \theta)\frac{d\hat{r}^i}{dt} + \theta \frac{d\hat{r}^{i+1}}{dt} \right] - \Delta t \, k_p \left[(1 - \theta)\hat{u}_w^i + \theta \hat{u}_w^{i+1} \right] = 0 \qquad (2.21)$$

which after substitution into eqn. 2.18 and rearrangement gives:

$$c^T \, \Delta \hat{r}^i - \Delta t \, k_p \left[\hat{u}_w^i + \theta \, \Delta \hat{u}_w^{\ i} \right] = 0 \qquad (2.22)$$

hence,

$$c^T \, \Delta \hat{r}^i - \Delta t \, \theta \, k_p \, \Delta \hat{u}_w^{\ i} = \Delta t \, k_p \, \hat{u}_w^i \qquad (2.23)$$

Equations 2.17 and 2.23 are combined in eqn. 2.24 to give the incremental coupled equations which form the basis of our finite element solutions:

$$\begin{bmatrix} k_m & c \\ c^T & -\Delta t \, \theta \, k_p \end{bmatrix} \left\{ \begin{array}{c} \Delta \hat{r}^i \\ \Delta \hat{u}_w^{\ i} \end{array} \right\} = \left\{ \begin{array}{c} \Delta f^i \\ \Delta t \, k_p \, \hat{u}_w^i \end{array} \right\} \qquad (2.24)$$

This is similar to the formulation reported by Sandhu (1981) and used by Griffiths *et al* (1991) .

Chapter 3

Elastic FE implementation

We now turn our attention to the actual finite element implementation of eqn. 2.24. A shorthand version of this equation at the element level can be written as:

$$k_e \Delta w^i = \Delta p^i \tag{3.1}$$

which after assembly is given an upper-case notation:

$$K_e \Delta W^i = \Delta P^i \tag{3.2}$$

in which ΔW^i holds all the incremental 'displacements' and ΔP^i hold all the incremental 'loads'.

Inspection of eqn. 3.2 indicates that for linear problems, the left hand side matrix K_e remains constant, thus the coefficient matrix needs to be factorised once only. The solution of ΔW^i at each time step therefore involves just the forward and back substitution phases. It should also be noted that K_e will have negative terms on its main diagonal due to the $-\Delta t \theta k_p$ terms. It will therefore not be possible to use a conventional Cholesky split approach which needs to take square roots of the diagonal terms. A conventional Gaussian elimination approach is recommended together with a 'skyline' strategy for storage.

A further saving can be made in the way in which the global right hand side vector is formed. The vector involves a combination of global load increments ΔF^i and products of the fluid 'stiffness' matrix K_p and excess pore pressures from the previous time U_w^i as shown in eqn. 3.3. Assembly of a global K_p matrix can be avoided by performing the product at the element level as will be shown. The decision as to whether to use an element–by–element approach depends to some extent on whether the user is seeking to economise on storage requirements.

$$\Delta \mathbf{P}^i = \left\{ \begin{array}{c} \Delta \mathbf{F}^i \\ \Delta t\, \mathbf{K}_p\, \mathbf{U_w}^i \end{array} \right\} \tag{3.3}$$

3.1   Example problem

The example used to illustrate the use of this program involves 1–d compression of a column of soil drained at its top surface only. The mesh shown in figure 3.1 involves four 8–node/4–node elements of total depth 10 units. Young's modulus and Poisson's ratio for the soil skeleton have been set to unity and zero respectively. The 'permeability' factors k_x/γ_w and k_y/γ_w have also both been set to unity.

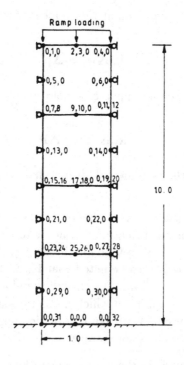

Figure 3.1: **Axially loaded example**

The mesh is subjected to axial 'ramp' loading as shown in figure 3.2 in which the axial stress builds up linearly to a maximum value of 1 unit at time t_o after which it remains constant. In this example, $t_o = 10.0$.

in order to non–dimensionalise the results a Time Factor is introduced whence:

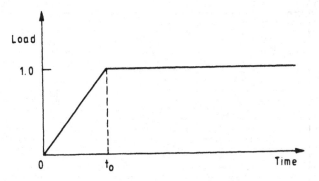

Figure 3.2: **Ramp loading**

$$T = \frac{c_v\, t}{D^2}$$

where D is the maximum drainage path (10.0 in this case), c_v is the coefficient of consolidation given by:

$$c_v = \frac{k}{m_v\, \gamma_w}$$

and m_v is the coefficient of volume compressibility given by:

$$m_v = \frac{(1 + \nu')(1 - 2\nu')}{E'(1 - \nu')}$$

For the particular properties chosen in this example,

$$c_v = m_v = 1.0$$

and

$$T = \frac{t}{100}$$

Figures 3.3 and 3.4 show respectively, the computed axial strain at the top of the column and the excess pore pressure at the base as a function of the Time Factor.

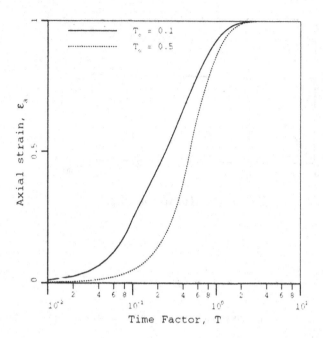

Figure 3.3: **Time Factor vs. Axial strain**

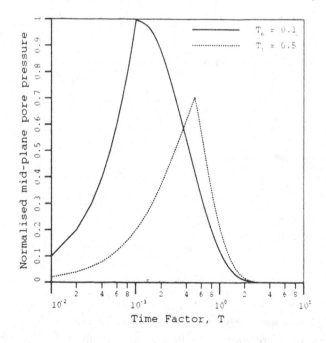

Figure 3.4: **Time Factor vs. Normalised base pore pressure**

For comparison, a second set of results corresponding to the case $T_o = 0.5$ is included. The results are in close agreement with Schiffman (1960) .

A logical choice for the time stepping parameter is to let $\theta = 0.5$, giving equivalence with the Trapezium Rule in which the derivatives at the end of each time step are equally weighted. This value of θ is unconditionally stable numerically, however the computational time step Δt still needs to be sufficiently small to avoid oscillatory results.

3.2   Program listing

```
      PROGRAM MP92
C
C        MODIFIED PROGRAM 9.2 BIOT CONSOLIDATION
C        ELASTIC SOLID IN PLANE STRAIN 8-NODE/4-NODE QUADRILATERALS
C        INCREMENTAL VERSION
C
C
C        ALTER NEXT LINE TO CHANGE PROBLEM SIZE
C
      PARAMETER (IKV=5000,ILOADS=200,INF=100,INX=10,INY=10,INO=20,
     +          NSTEPS=500,INP=30)
C
      REAL WIDTH(INX),DEPTH(INY),KV(IKV),VAL(INO),LOADS(ILOADS),
     +     ANS(ILOADS),RT(INP),RL(INP),AL(NSTEPS),SX(INX,INY,4),
     +     SY(INX,INY,4),TXY(INX,INY,4)
      INTEGER NF(INF,3),NO(INO)
C
      REAL DEE(3,3),SAMP(3,2),COORD(8,2),DERIVT(8,2),JAC(2,2),
     +     JAC1(2,2),KAY(2,2),DER(2,8),DERIV(2,8),KDERIV(2,8),BEE(3,16),
     +     DBEE(3,16),BT(16,3),BTDB(16,16),KM(16,16),DTKD(4,4),KP(4,4),
     +     KE(20,20),FUN(8),C(16,4),VOLF(16,4),VOL(16),PHILO(4),FUNF(4),
     +     COORDF(4,2),DERF(2,4),DERIVF(2,4),PHIO(4)
      INTEGER G(20)
C
      DATA IJAC,IJAC1,IKAY,IDER,IDERIV,IKDERV,IT,IDERF,IDERVF/9*2/
      DATA IDEE,ISAMP,IBEE,IDBEE,NODOF,IH/6*3/
      DATA ICORDF,IDTKD,IKP,NODF/4*4/,IKE,IKD,ITOT/3*20/
      DATA ICOORD,IDERVT,NOD/3*8/,IBT,IBTDB,IKM,IC,IVOLF,IDOF/6*16/
C
C        INPUT AND INITIALISATION
```

```
C
      READ (5,*) NXE,NYE,N,IW,NN,NR,NGP,PERMX,PERMY,E,V,DTIM,ISTEP,THETA
      READ (5,*) (WIDTH(I),I=1,NXE+1)
      READ (5,*) (DEPTH(I),I=1,NYE+1)
      CMV = (1.+V)* (1.-2.*V)/E/ (1.-V)
      CV = PERMX/CMV
      TFAC = CV/ (DEPTH(NYE+1)*DEPTH(NYE+1))
      CALL READNF(NF,INF,NN,NODOF,NR)
C
C    READ LOAD WEIGHTINGS
C
      READ (5,*) NL, (NO(I),VAL(I),I=1,NL)
C
C    READ NO. OF POINTS IN LOAD/TIME HISTORY (NP)
C    READ TIME (RT) AND LOAD (RL) FOR LOAD/TIME HISTORY
C
      READ (5,*) NP
      DO 10 I = 1,NP
   10 READ (5,*) RT(I),RL(I)
      CALL LFUNC(RT,RL,NP,DTIM,AL,NTP)
C
C    READ FREEDOM NUMBER AT WHICH OUTPUT IS REQUIRED
C
      READ (5,*) NFOUT
      IR = N* (IW+1)
      CALL NULVEC(ANS,N)
      CALL NULVEC(KV,IR)
      CALL FMDEPS(DEE,IDEE,E,V)
      CALL NULL(KAY,IKAY,IT,IT)
      KAY(1,1) = PERMX
      KAY(2,2) = PERMY
      CALL GAUSS(SAMP,ISAMP,NGP)
C
C    ELEMENT MATRIX INTEGRATION AND ASSEMBLY
C
      DO 20 IP = 1,NXE
         DO 20 IQ = 1,NYE
            CALL GEOUVP(IP,IQ,NXE,WIDTH,DEPTH,COORD,ICOORD,COORDF,
     +                  ICORDF,G,NF,INF)
            CALL NULL(KM,IKM,IDOF,IDOF)
            CALL NULL(C,IC,IDOF,NODF)
            CALL NULL(KP,IKP,NODF,NODF)
```

```
          IG = 0
          DO 30 I = 1,NGP
             DO 30 J = 1,NGP
                IG = IG + 1
                SX(IP,IQ,IG) = 0.
                SY(IP,IQ,IG) = 0.
                TXY(IP,IQ,IG) = 0.
                CALL FMQUAD(DER,IDER,FUN,SAMP,ISAMP,I,J)
                CALL MATMUL(DER,IDER,COORD,ICOORD,JAC,IJAC,IT,NOD,
     +                     IT)
                CALL TWOBY2(JAC,IJAC,JAC1,IJAC1,DET)
                CALL MATMUL(JAC1,IJAC1,DER,IDER,DERIV,IDERIV,IT,
     +                     IT,NOD)
                CALL NULL(BEE,IBEE,IH,IDOF)
                CALL FORMB(BEE,IBEE,DERIV,IDERIV,NOD)
                CALL VOL2D(BEE,IBEE,VOL,NOD)
                CALL MATMUL(DEE,IDEE,BEE,IBEE,DBEE,IDBEE,IH,IH,
     +                     IDOF)
                CALL MATRAN(BT,IBT,BEE,IBEE,IH,IDOF)
                CALL MATMUL(BT,IBT,DBEE,IDBEE,BTDB,IBTDB,IDOF,IH,
     +                     IDOF)
                QUOT = DET*SAMP(I,2)*SAMP(J,2)
                CALL MSMULT(BTDB,IBTDB,QUOT,IDOF,IDOF)
                CALL MATADD(KM,IKM,BTDB,IBTDB,IDOF,IDOF)
C
C     FLUID CONTRIBUTION
C
                CALL FORMLN(DERF,IDERF,FUNF,SAMP,ISAMP,I,J)
                CALL MATMUL(DERF,IDERF,COORDF,ICORDF,JAC,IJAC,IT,
     +                     NODF,IT)
                CALL TWOBY2(JAC,IJAC,JAC1,IJAC1,DET)
                CALL MATMUL(JAC1,IJAC1,DERF,IDERF,DERIVF,IDERVF,
     +                     IT,IT,NODF)
                CALL MATMUL(KAY,IKAY,DERIVF,IDERVF,KDERIV,IKDERV,
     +                     IT,IT,NODF)
                CALL MATRAN(DERIVT,IDERVT,DERIVF,IDERVF,IT,NODF)
                CALL MATMUL(DERIVT,IDERVT,KDERIV,IKDERV,DTKD,
     +                     IDTKD,NODF,IT,NODF)
                QUOT = DET*SAMP(I,2)*SAMP(J,2)
                PROD = QUOT*DTIM
                CALL MSMULT(DTKD,IDTKD,PROD,NODF,NODF)
                CALL MATADD(KP,IKP,DTKD,IDTKD,NODF,NODF)
```

```
                        CALL VVMULT(VOL,FUNF,VOLF,IVOLF,IDOF,NODF)
                        CALL MSMULT(VOLF,IVOLF,QUOT,IDOF,NODF)
                        CALL MATADD(C,IC,VOLF,IVOLF,IDOF,NODF)
     30         CONTINUE
                CALL FORMKE(KM,IKM,KP,IKP,C,IC,KE,IKE,IDOF,NODF,THETA)
                CALL FORMKV(KV,KE,IKE,G,N,ITOT)
   20 CONTINUE
C
C     REDUCE LEFT HAND SIDE
C
      CALL BANRED(KV,N,IW)
C
C     TIME STEPPING LOOP
C
      WRITE (6,'(2E12.4)') 0.,ANS(NFOUT)
      DO 40 NS = 1,ISTEP
          CALL NULVEC(LOADS,N)
          DO 50 IP = 1,NXE
              DO 50 IQ = 1,NYE
                  CALL GEOUVP(IP,IQ,NXE,WIDTH,DEPTH,COORD,ICOORD,COORDF,
     +                        ICORDF,G,NF,INF)
                  CALL NULL(KP,IKP,NODF,NODF)
                  DO 60 I = 1,NGP
                      DO 60 J = 1,NGP
                          CALL FORMLN(DERF,IDERF,FUNF,SAMP,ISAMP,I,J)
                          CALL MATMUL(DERF,IDERF,COORDF,ICORDF,JAC,IJAC,
     +                                IT,NODF,IT)
                          CALL TWOBY2(JAC,IJAC,JAC1,IJAC1,DET)
                          CALL MATMUL(JAC1,IJAC1,DERF,IDERF,DERIVF,
     +                                IDERVF,IT,IT,NODF)
                          CALL MATMUL(KAY,IKAY,DERIVF,IDERVF,KDERIV,
     +                                IKDERV,IT,IT,NODF)
                          CALL MATRAN(DERIVT,IDERVT,DERIVF,IDERVF,IT,
     +                                NODF)
                          CALL MATMUL(DERIVT,IDERVT,KDERIV,IKDERV,DTKD,
     +                                IDTKD,NODF,IT,NODF)
                          QUOT = DET*SAMP(I,2)*SAMP(J,2)*DTIM
                          CALL MSMULT(DTKD,IDTKD,QUOT,NODF,NODF)
                          CALL MATADD(KP,IKP,DTKD,IDTKD,NODF,NODF)
     60               CONTINUE
                  DO 70 M = IDOF + 1,ITOT
                      IF (G(M).EQ.0) PHI0(M-IDOF) = 0.
```

```
 70                   IF (G(M).NE.0) PHIO(M-IDOF) = ANS(G(M))
                      CALL MVMULT(KP,IKP,PHIO,NODF,NODF,PHILO)
                      DO 80 M = IDOF + 1,ITOT
                         IF (G(M).EQ.0) GO TO 80
                         LOADS(G(M)) = LOADS(G(M)) + PHILO(M-IDOF)
 80                   CONTINUE
 50         CONTINUE
C
C     RAMP LOADING
C
            DO 90 I = 1,NL
 90         LOADS(NO(I)) = VAL(I)*AL(NS)
            CALL BACSUB(KV,LOADS,N,IW)
            CALL VECADD(ANS,LOADS,ANS,N)
            WRITE (6,'(2E12.4)') NS*DTIM*TFAC,-ANS(NFOUT)
 40     CONTINUE
        STOP
        END
C
        SUBROUTINE LFUNC(RT,RL,NP,DTIM,AL,NTP)
C
C        THIS SUBROUTINE FORMS THE LOAD (OR DISPLACEMENT)-TIME
C        FUNCTION AT INTERVALS OF THE CALCULATION TIME STEP DTIM
C
        REAL RT(*),RL(*),AL(*)
        NTP = NINT(RT(NP)/DTIM)
        AOLD = RL(1)
        DT = RT(1)
        J = 2
        DO 10 I = 1,NTP
           DT = DT + DTIM
 20        IF (DT/RT(J).LT.1.0000001) THEN
               ANEW = RL(J-1) + (DT-RT(J-1))* (RL(J)-RL(J-1))/
      +              (RT(J)-RT(J-1))
               AL(I) = ANEW - AOLD
               AOLD = ANEW
           ELSE
               J = J + 1
               GO TO 20
           END IF
 10     CONTINUE
        RETURN
```

```
          END
C
          SUBROUTINE FORMKE(KM,IKM,KP,IKP,C,IC,KE,IKE,IDOF,NODF,THETA)
C
C         THIS SUBROUTINE FORMS THE ELEMENT COUPLED STIFFNESS
C         MATRICES KE FROM THE ELASTIC STIFFNESS KM,
C         THE FLUID 'STIFFNESS' KP AND COUPLING MATRIX C
C
          REAL KM(IKM,*),KP(IKP,*),C(IC,*),KE(IKE,*)
          DO 10 I = 1,IDOF
              DO 20 J = 1,IDOF
   20         KE(I,J) = KM(I,J)
              DO 30 K = 1,NODF
                  KE(I,IDOF+K) = C(I,K)
   30         KE(IDOF+K,I) = C(I,K)
   10 CONTINUE
          DO 40 I = 1,NODF
              DO 40 K = 1,NODF
   40 KE(IDOF+I,IDOF+K) = -KP(I,K)*THETA
          RETURN
          END
```

3.3 Notes on the program

The listing consists of a main program MP92 followed by two subroutines, LFUNC and
FORMKE. The program is very similar to Program 9.2 published by Smith and Griffiths
(1988) , the main differences being that the present approach uses the incremental
formulation described previously. All the other subroutines are in the Appendix of
the Smith and Griffiths textbook. Some other important refinements are now
highlighted here.

1. The program allows any load/time function to be applied to the mesh. This is
   read in at do–loop 10 as a sequence of (load, time) coordinates. The program
   automatically assumes linear interpolation between the coordinates at the
   calculation time step using subroutine LFUNC.

2. Do–loop 20 scans each element and uses 2×2 Gaussian numerical integration
   to form the k_m, c and k_p matrices. Subroutine FORMKE combines these matrices
   to form the element matrix k_e.

3. The 'loading' terms due to the excess pore pressures are formed on an *element*
   by element basis in do–loop 50. In this loop, the element matrix k_p is reformed

and the appropriate excess pore pressure terms extracted from the global 'displacements' vector ANS.

4. The time stepping do–loop 40 ends with a print out of the Time Factor and the required freedom. The section in which the strains and effective stresses are computed have been removed for the sake of brevity, however these can be put back by the user if required.

The code is transparent and modular, enabling users to make changes in any way which may suit their own particular problem. A good example would be in the conversion of the program to three–dimensions. This could be achieved relatively easily by replacing certain subroutines by their 3-d counterparts.

For problems with more complicated boundaries, the user must replace the geometry subroutine GEOUVP which is only suitable for rectangular geometries, by some other means of generating the nodal coordinates COORD and the 'steering' vector G. A number of proprietary codes are available for this kind of pre-processing.

3.4   Sample data

Much of the data is self explanatory, however for a detailed description of the
meaning of variable names the reader is again referred to the Smith and Griffiths
(1988) text. Although the problem under consideration is exactly the same as that
described under Program 9.2, the data here differs in the 'Loading data', which in
this presentation includes 'Nodal weightings' and a 'Load/time function'. An
additional data point relating to the output is also included. In this particular case,
the computed load–time history at freedom number 31, the base pore pressure, is
requested.

```
                         Mesh Data
                NXE  NYE   N   IW  NN  NR  NGP
                 1    4   32  13  23  23   2

                     Element data
                 PERMX  PERMY   E   V
                   1.     1.    1.  0.

                 Time integration data
                   DTIM  ISTEP  THETA
                    1.    300    0.5

            Geometry data -- WIDTH , DEPTH
                  0.   1.
                  0.  -2.5  -5.  -7.5  -10.

               Node freedom data -- READNF
    1  0  1  0     2  1  1  0     3  0  1  0     4  0  1  0      5  0  1  0
    6  0  1  1     7  1  1  0     8  0  1  1     9  0  1  0     10  0  1  0
   11  0  1  1    12  1  1  0    13  0  1  1    14  0  1  0     15  0  1  0
   16  0  1  1    17  1  1  0    18  0  1  1    19  0  1  0     20  0  1  0
   21  0  0  1    22  0  0  0    23  0  0  1
```

3.5   Sample output

The following output was produced by the listed program and data. The two columns of results represent the Time Factor and (in this case) the base pore pressure. The output is deliberately designed to be very basic, since this minimises the length of the main program and encourages users to customise the program to suit their particular requirements.

Time Factor	Freedom no. 31
0.0000E-00	0.0000E-00
0.1000E-01	0.1999E-01
0.2000E-01	0.4004E-01
0.3000E-01	0.5994E-01
0.4000E-01	0.7994E-01
0.5000E-01	0.1000E+00
0.6000E-01	0.1202E+00
0.7000E-01	0.1403E+00
0.8000E-01	0.1602E+00
0.9000E-01	0.1799E+00
⋮	⋮
0.2910E+01	0.1782E-02
0.2920E+01	0.1738E-02
0.2930E+01	0.1695E-02
0.2940E+01	0.1651E-02
0.2950E+01	0.1613E-02
0.2960E+01	0.1573E-02
0.2970E+01	0.1534E-02
0.2980E+01	0.1496E-02
0.2990E+01	0.1459E-02
0.3000E+01	0.1423E-02

Chapter 4

Failure criteria for geomaterials

In this chapter, simple nonlinear models for soil are introduced and the properties of some of the best known failure criteria for soil are discussed. This section is included because the elastic model used so far has considerable limitations in the modelling of geomaterials. Although the elastic model may be justified for medium dense materials at working stress levels, it becomes highly unsatisfactory if nonlinear stress–strain behaviour is anticipated, or if there is significant volume change tendency (contractive or dilative) during shear.

Volume change tendency in geomaterials becomes very important in the presence of pore pressures. In the extreme case of undrained conditions, volume change tendency results in pore pressure changes which immediately influence effective stresses and hence the stiffness and strength of frictional materials. Thus a dilative material may appear to have greatly increased strength under undrained (or even partially drained) conditions, whereas a contractive soil may suffer dramatic loss of strength.

The scope of this presentation does not allow an in–depth study of the constitutive behaviour of geomaterials, so for the purposes of introducing nonlinearity into the Biot formulation, a simple 5–parameter elastic–perfectly plastic model will be considered as summarised in Table 4.1.

Table 4.1: **Five–parameter model**

E'	Young's modulus
ν'	Poisson's ratio
ϕ'	Friction angle
c'	Cohesion
ψ	Dilation angle

Figure 4.1 indicates that this model consists of a linear section followed by a failure section. The volume change predicted by the model is also shown in the figure, together with the behaviour that might be observed in a 'real' soil, shown as a dotted line.

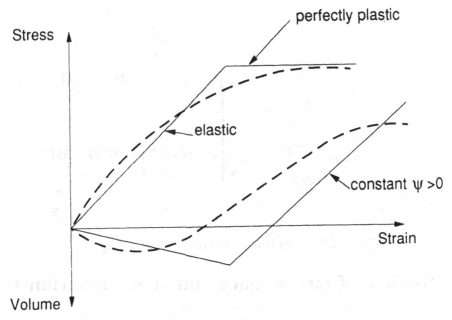

Figure 4.1: **Elastic–perfectly plastic model**

By choosing an 'average' stiffness for the elastic part of the model and a reliable failure criterion, this model can give reasonable predictions of both deformations and collapse loads in drained boundary value problems. The limitations of the model are more pronounced under undrained or partially drained conditions since it is a poor predictor of volume change, the simplest version of the model being based on a constant dilation angle ψ. The undrained effective stress paths shown in figure 4.2 for a range of densities indicate that the simple model will perform adequately for 'medium–dense' materials.

Even the most sophisticated constitutive models for soil must include some form of failure criterion which determines the limiting strength of the material for all possible stress paths. A study of some of the better-known criteria is included here because some of them can give surprisingly poor representations of soil strength for certain stress paths.

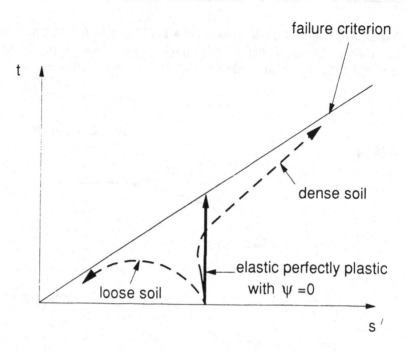

Figure 4.2: **Undrained effective stress paths**

4.1 Review of stress space and stress invariants

In order to study the general stress fields that occur in a complicated boundary value problem, it is convenient to use principal stress space, which also leads to a convenient geometric representation of various failure criteria. A stress point in principal stress space can be defined using the following invariants:

$$(s, t, \theta) \qquad (4.1)$$

where

$$s = \frac{1}{\sqrt{3}}(\sigma_1 + \sigma_2 + \sigma_3) \qquad (4.2)$$

$$t = \frac{1}{\sqrt{3}}\left[(\sigma_1 - \sigma_2)^2 + (\sigma_2 - \sigma_3)^2 + (\sigma_3 - \sigma_1)^2\right]^{\frac{1}{2}} \qquad (4.3)$$

$$\theta = \arctan\left[\frac{1}{\sqrt{3}} \frac{(\sigma_1 - 2\sigma_2 + \sigma_3)}{(\sigma_1 - \sigma_3)}\right] \tag{4.4}$$

From equations 4.1– 4.4, s represents the perpendicular distance of the deviatoric plane containing the stress point from the origin, t represents the radial distance of the stress point from the space diagonal, and θ is the Lode angle giving an angular measure of position within the deviatoric plane. In this context, the Lode angle varies in the range:

$$-30° \leq \theta \leq 30° \tag{4.5}$$

where $\theta = -30°$ corresponds to triaxial extension, and $\theta = 30°$ to triaxial compression. It is often useful to non-dimensionalise principal stress space as shown in figure 4.3, by dividing all lengths by the invariant s leading to normalised coordinates:

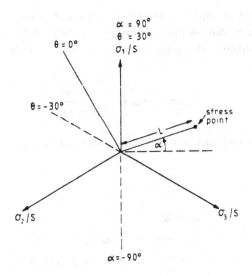

Figure 4.3: **Dimensionless principal stress space**

$$(1, L, \theta) \tag{4.6}$$

where

$$L = \frac{t}{s} \tag{4.7}$$

This overcomes any ambiguity over the sign convention, as it is recognised that the majority of geotechnical engineers use a compression- positive convention. Other invariants that will be used later are the stress invariants I_1, I_2, I_3 and for completeness these are defined below:

$$I_1 = \sigma_1 + \sigma_2 + \sigma_3$$

$$I_2 = \sigma_1\sigma_2 + \sigma_2\sigma_3 + \sigma_3\sigma_1 \tag{4.8}$$

$$I_3 = \sigma_1\sigma_2\sigma_3$$

4.2   The Mohr-Coulomb angle in relation to principal stress space

The equivalent Mohr-Coulomb friction angle ϕ'_{mc} corresponding to a particular location in stress space, can be found from invariants L and θ. Due to the corners on the Mohr-Coulomb surface, the general expression can be written in the form:

$$\phi'_{mc} = \arcsin\left[\frac{\sqrt{3}L\cos\theta}{\sqrt{2} + L\sin\theta}\right] \tag{4.9}$$

where, with reference to figure 4.3 and taking account of symmetry about the σ_1/s-axis:

$$\begin{aligned}
\theta &= \alpha + 60° \quad \text{if} \quad -90° \leq \alpha < -30° \\
\theta &= -\alpha \quad \text{if} \quad -30° \leq \alpha < +30° \\
\theta &= \alpha - 60° \quad \text{if} \quad +30° \leq \alpha \leq +90°
\end{aligned} \tag{4.10}$$

4.3   Circular conical criteria

These were among the earliest surfaces (Drucker and Prager 1952, Drucker *et al* 1957) suggested as being suitable for representing the strength of soils. They project as circles in deviatoric planes, and are usually fitted exactly to Mohr-Coulomb's hexagon at certain locations on their circumference. Many circular fits are possible (Humpheson 1976, Zienkiewicz *et al* 1978) , but the two considered here are the limiting 'Internal' and 'External' cones (Griffiths 1986) .

4.3.1   External Cone

Referring to figure 4.4, the External cone is fitted to the Mohr-Coulomb criterion at the apexes of the hexagon corresponding to triaxial compression. When viewed in a deviatoric plane the circle is centred at the origin with a radius given by:

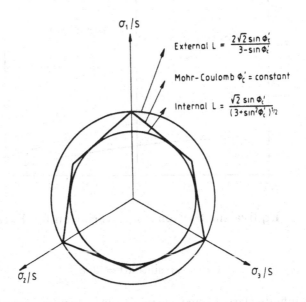

Figure 4.4: **Mohr–Coulomb with External and Internal cones**

$$L = \frac{2\sqrt{2}\sin\phi'_c}{3 - \sin\phi'_c} \tag{4.11}$$

where ϕ'_c is the friction angle of the circumscribed Mohr-Coulomb surface. Substituting L into eqn. 4.9 gives the following expression relating the equivalent friction angle ϕ'_{mc} to the angular invariant θ:

$$\phi'_{mc} = \arcsin\left[\frac{2\sqrt{3}\sin\phi'_c\cos\theta}{3 + \sin\phi'_c(2\sin\theta - 1)}\right] \tag{4.12}$$

Due to the six-fold symmetry in stress space, this relationship is illustrated in figure 4.5 over a 60° sector of principal stress space for three External cones fitted to $\phi'_c = 25°, 30°$ and 36.87°. As expected $\phi'_{mc} = \phi'_c$ for triaxial compression stress paths ($\theta = 30°$), however considerable overestimates of strength can occur for other stress paths. For example, a cone fitted at $\phi'_c = 30°$ could predict an equivalent friction angle ϕ'_{mc} as high as $49.11°$ for a stress path given by $\theta = -23.58°$, falling only slightly to $48.59°$ in triaxial extension ($\theta = -30°$). The well-known singularity (Bishop 1966) of this surface is also shown in figure 4.5, which occurs when the cone is fitted to a triaxial compression friction angle given by:

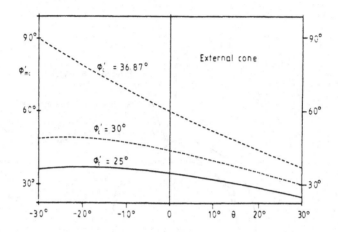

Figure 4.5: **Equivalent friction angle variation for External cones**

$$\phi'_c = \arcsin(0.6) = 36.87°$$ (4.13)

In this case an infinite shear strength is predicted in triaxial extension as indicated by $\phi'_{mc} = 90°$.

The External cone is clearly an unacceptable failure surface for frictional soil.

4.3.2   Internal Cone

Referring to figure 4.4, the Internal cone is internally tangent, and hence represents a lower-bound circular fit to the Mohr-Coulomb surface. This surface will clearly predict equivalent friction angles that are less than or equal to the that of the surrounding Mohr-Coulomb surface for all possible stress paths. The radius of the Internal cone as viewed in a deviatoric plane is given by:

$$L = \frac{\sqrt{2}\sin\phi'_c}{(3 + \sin^2\phi'_c)^{\frac{1}{2}}}$$ (4.14)

where ϕ'_c is the friction angle of the surrounding Mohr-Coulomb surface. The point of tangency of the two surfaces is readily shown (Griffiths 1986) to occur at the angular location given by :

$$\theta = \arctan\left[\frac{-\sin \phi_c'}{\sqrt{3}}\right] \tag{4.15}$$

Substituting L into eqn. 4.9 gives the following expression relating the equivalent friction angle ϕ_{mc}' to the angular invariant θ :

$$\phi_{mc}' = \arcsin\left[\frac{\sqrt{3}\sin \phi_c' \cos \theta}{(3 + \sin^2 \phi_c')^{\frac{1}{2}} + \sin \phi_c' \sin \theta}\right] \tag{4.16}$$

Due to the six-fold symmetry, in stress space, this relationship is illustrated in figure 4.6 over a 60° sector for three Internal cones fitted to $\phi_c' = 20^\circ, 30^\circ$ and 40° . As expected, the maximum value of ϕ_{mc}' occurs at a value of θ given by eqn. 4.15. Although the Internal cone will always give conservative estimates of soil strength, certain stress paths produce unacceptably pessimistic predictions. For example, a cone internally tangent to Mohr-Coulomb with $\phi_c' = 40^\circ$ would predict an equivalent friction angle as low as $\phi_{mc}' = 26.39^\circ$ in triaxial compression ($\theta = 30^\circ$).

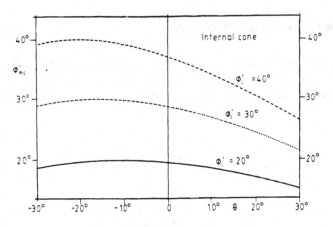

Figure 4.6: **Equivalent friction angle variation for Internal cones**

4.4   Smooth approximations to Mohr-Coulomb

Although simple geometrically, the circular cones have been shown to give very poor representations of soil strength. A more rational approach is to find a failure criterion without corners, and based on shear strength data obtained from a true triaxial device. Two such surfaces are now considered as shown in figure 4.7, both of which are supported by experimental evidence produced by their authors.

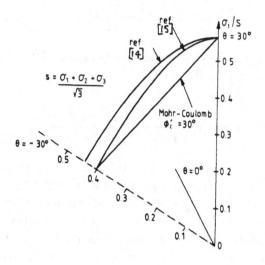

Figure 4.7: **Lade/Duncan and Matsuoka/Nakai in relation to Mohr–Coulomb**

4.4.1   Lade failure criterion

This surface (Lade and Duncan 1975) is seen to coincide with Mohr-Coulomb in triaxial compression, but to run outside the hexagon for all other stress paths. The criterion is expressed in invariant form as follows-

$$\frac{I_1^3}{I_3} = K_L \tag{4.17}$$

where

$$K_L = \frac{(3 - \sin \phi_c')^3}{(1 + \sin \phi_c')(1 - \sin \phi_c')^2} \tag{4.18}$$

4.4.2   Matsuoka/Nakai failure criterion

This surface (Matsuoka and Nakai 1974) is seen to coincide with Mohr-Coulomb at all apexes corresponding to triaxial extension and compression. The relationship between this surface and Mohr-Coulomb is analogous to that between von-Mises and Tresca. The criterion is expressed in invariant form as follows:

$$\frac{I_1 I_2}{I_3} = K_{MN} \tag{4.19}$$

where

$$K_{MN} = \frac{9 - \sin^2 \phi'_c}{1 - \sin^2 \phi'_c} \tag{4.20}$$

Not only is this surface more conservative than Lade's for all stress paths, but the maximum value of ϕ'_{mc} occurs at a different location. In the Lade surface, the maximum always occurs at $\theta = 0^\circ$, whereas in the Matsuoka-Nakai surface, the maximum depends on ϕ'_c. For comparison, both surfaces corresponding to $\phi'_c = 40^\circ$ are shown together in figure 4.8. It can be noted from this figure that even these 'sophisticated' surfaces predict equivalent Mohr-Coulomb friction angles that differ by as much as 6° in triaxial extension. To put this in perspective, under plane strain conditions a change in the friction angle of 4° from 44° to 48° causes the the predicted bearing capacity of a strip footing to increase by a factor of three (see e.g Lambe and Whitman 1969) !

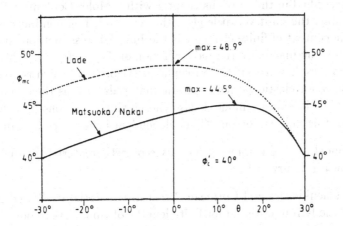

Figure 4.8: **Lade/Duncan and Matsuoka/Nakai fitted at** $\phi'_c = 40^\circ$

The Mohr-Coulomb surface remains attractive on the grounds that it is conservative and 'familiar'. As a conservative smooth alternative however, the Matsuoka–Nakai surface can be recommended.

Chapter 5

Elasto–plastic FE implementation

In this section a program is presented in which a simple elasto-plastic constitutive model is implemented with the Biot coupled equations. This represents a natural progression from the program presented in the previous chapter which was limited to elastic materials. Once we know how to incorporate simple nonlinear constitutive models into the Biot framework, the implementation of more advanced models presents relatively little additional complexity.

5.1   Viscoplasticity

The simple model to be adopted in this section is the 5-parameter elastic–perfectly plastic model described in the previous chapter with a Mohr-Coulomb failure criterion. Arguably, the most versatile algorithm for modelling nonlinear material behaviour in the context of finite elements is the Initial-Stress method, since it incorporates a 'plastic matrix' in the redistribution of illegal stresses. The method that will be used here however, is the Viscoplastic Algorithm, which is known to give robust convergence for elastic–perfectly plastic materials such as the one under consideration. The viscoplastic model is also chosen since it is consistent with the general theme of this publication on, "Time dependent behaviour of geomaterials".

A rheological model for 1-d compression of a viscoplastic material with pore pressures is shown in figure 5.1.

In this method (Zienkiewicz and Cormeau, 1974), the material is allowed to sustain stresses outside the failure criterion for finite lengths of time. Overshoot of the failure criterion as signified by a positive value of the failure function F is an integral part of the algorithm and is actually used to drive the stress redistribution process. The rate at which plastic strains are generated is related to the amout by which the failure

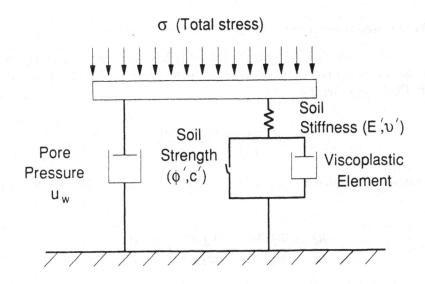

Figure 5.1: **Rheological model of 1-d viscoplastic/Biot material**

function has been violated. A simple way of expressing this is through the equation:

$$\dot{\epsilon}^{vp} = F\frac{\partial Q}{\partial \sigma} \tag{5.1}$$

where F is the failure function, Mohr–Coulomb in this case, and Q is the potential function which is geometrically similar to Mohr–Coulomb but with the friction angle ϕ replaced by a dilation angle ψ. It should be noted that eqn. 5.1 implies a pseudo–viscosity which has no basis in real time. The viscoplastic algorithm when used in this way can be considered as a numerical convenience for redistributing illegal stresses. If realistic data was available on the actual viscous behaviour of soil, it could be easily incorporated into the viscoplastic framework described.

Multiplication of the viscoplastic strain–rate from eqn. 5.1 by a pseudo time step gives an increment of plastic strain which can be accumulated from one time step to the next; thus

$$(\delta\epsilon^{vp})^{i} = \Delta t(\dot{\epsilon}^{vp})^{i} \tag{5.2}$$

and

$$(\Delta \epsilon^{vp})^i = (\Delta \epsilon^{vp})^{i-1} + (\delta \epsilon^{vp})^i \tag{5.3}$$

where the superscript refers to the time step or iteration number.

The time step for unconditional numerical stability has been shown by Cormeau (1975) to be dependent on the failure criterion under consideration. For example with the Mohr–Coulomb criterion:

$$\Delta t = \frac{4(1 + \nu)(1 - 2\nu)}{E(1 - 2\nu + \sin^2 \phi)} \tag{5.4}$$

The derivatives of the plastic potential function Q are expressed through the Chain Rule as:

$$\frac{\partial Q}{\partial \sigma} = \frac{\partial Q}{\partial \sigma_m} \frac{\partial \sigma_m}{\partial \sigma} + \frac{\partial Q}{\partial J_2} \frac{\partial J_2}{\partial \sigma} + \frac{\partial Q}{\partial J_3} \frac{\partial J_3}{\partial \sigma} \tag{5.5}$$

where σ_m, J_2 and J_3 are the mean stress and the second and third deviatoric stress invariants respectively.

The 'body–loads' $\mathbf{P}$ used to redistribute illegal stresses are accumulated at each time step by evaluating the integral given by eqn. 5.6 over any element which contains a 'yielding' Gauss–point.

$$\mathbf{P}^i = \mathbf{P}^{i-1} + \int_{V^e} \mathbf{B}^T \mathbf{D}^e (\delta \epsilon^{vp})^i \, \mathrm{d}(\mathrm{vol}) \tag{5.6}$$

This process is repeated until all Gauss–point stresses have been returned to the failure surface within certain tolerances. The convergence criterion is satisfied when the change in nodal displacements from one iteration to the next is less than a user–defined value.

A potential source of confusion in the Viscoplastic/Biot algorithm is that there are two layers of time steps taking place simultaneously. The time step associated with the Biot process can be considered to be 'real' time, whereas the time step associated with the viscoplastic iterative process can be considered to be a 'fictitious' time. At each Biot time step, the viscoplastic algorithm takes a certain number of viscoplastic time steps to reach convergence.

A more detailed description of the whole algorithm can be found in Griffiths (1980) .

5.2   Example problem

The example used to illustrate the use of this program involves axial compression of a square block of frictional material. The mesh shown in figure 5.2 models one quarter of the problem which is a unit square.

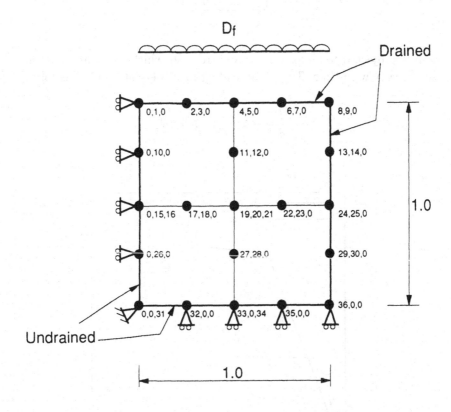

Figure 5.2: **Axial compression example**

The two axes of symmetry are no–flow boundaries and the other two sides are free–draining. The mesh uses four 8–node/4–node elements. The five–parameter soil model has the properties shown in Table 5.1.

The 'permeability' factors k_x/γ_w and k_y/γ_w are both set to unity and the initial effective stresses at all Gauss–points within the mesh are set to the following values:

$$\sigma_x' = \sigma_y' = \sigma_z' = -10.0$$

$$\tau_{xy} = 0.0$$

Table 5.1: **Soil Properties**

E'	10000.0
ν'	0.25
ϕ'	30°
c'	0.0
ψ	0.0

The mesh is subjected to an axial deviator stress D_f which starts at zero and increases linearly with time. Two axial loading rates were attempted as follows:

$$\frac{dD_f}{dt} = 5 \times 10^3, \; 5 \times 10^6$$

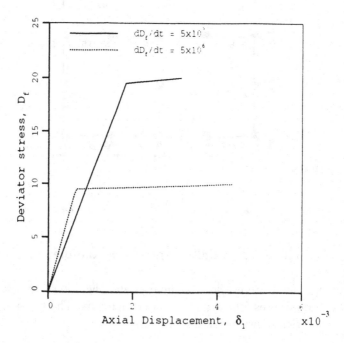

Figure 5.3: **Computed response for two different loading rates**

Figure 5.3 shows plots of the deviator stress vs. axial displacement for both these cases. The collapse load for the faster loading rate is lower than for the slower loading rate. These results correspond to undrained and drained conditions respectively and the computed deviator stress a failure in both cases is in close agreement with the analytical solution derived by Griffiths (1985) which gives:

$$\frac{D_f}{\sigma_3} = \frac{(K_p - 1)(2\beta_{ps} + 1)}{(K_p + 1)\beta_{ps} + 1} \tag{5.7}$$

where $\beta_{ps} \longrightarrow 0$ for drained conditions and $\beta_{ps} \longrightarrow \infty$ for undrained conditions.

In the present example,

$$\phi' = 30°, \text{ and } K_p = 3$$

hence the expected results for fully drained and undrained behaviour are,

$$D_f/\sigma_3 = 2 \text{ and } 1 \text{ respectively.}$$

It should be emphasised that the undrained solution is based on the assumption that there is no plastic volume change during yield (see figure 4.2). In the present example this condition has applied since the dilation angle ψ was set to zero.

5.3   Program listing

```
      PROGRAM MCPROG
C
C         MODIFIED PROGRAM 9.2 BIOT CONSOLIDATION
C         ELASTO-PLASTIC MOHR-COULOMB SOLID
C         PLANE STRAIN 8-NODE/4-NODE QUADRILATERALS
C         INCREMENTAL VERSION
C
      PARAMETER (INO=100,INF=300,NSTEPS=1000,INX=20,INY=20,IFLU=150,
     +           ILOADS=800,IBK=210000,IBP=1500,IEL=50,IEVPT=1000)
C
      INTEGER NF(INF,3),NF2(INF,2),NF3(INF),LVEC(INF),NO(INO)
      INTEGER NITS(NSTEPS)
      INTEGER G(20),G3(4)
C
      REAL POGH2(NSTEPS),XOL(NSTEPS),FM(NSTEPS),STORKB(INO),WIDTH(INX),
     +     DEPTH(INY),FLUID(IFLU),ANS2(IFLU),DISPL(ILOADS),
     +     LOADS(ILOADS),BDYLDS(ILOADS),OLDIS(ILOADS),ANS(ILOADS),
```

```
   +       BK(IBK),BP(IBP),PORE(2,2,IEL),VAL(INO),SX(INX,INY,4),
   +       SY(INX,INY,4),SZ(INX,INY,4),TXY(INX,INY,4),EVPT(IEVPT)
C
      REAL STRESS(4),GC(2),ELD(16),VOL(16),EPS(4),SIGMA(4),ERATE(4),
   +       EVP(4),DEVP(4),FUN(8),ELOAD(16),BLOAD(16),FUNF(4)
C
      REAL M1(4,4),M2(4,4),M3(4,4),DEE(4,4),SAMP(3,2),COORD(8,2),
   +       DBEE(4,16),FLOW(4,4),JAC(2,2),JAC1(2,2),DER(2,8),DERIV(2,8),
   +       BEE(4,16),BTDB(16,16),KM(16,16),BT(16,4),DERIVT(8,2),
   +       KAY(2,2),KDERIV(2,4),DTKD(4,4),KP(4,4),KE(20,20),COORDF(4,2),
   +       DERF(2,4),DERIVF(2,4),C(16,4),VOLF(16,4)
C
      DATA IJAC,IJAC1,IDER,IDERIV,IKAY,IKDERV,IDERF,IDERVF,IT/9*2/
      DATA ISAMP,NODOF/2*3/
      DATA IPL,IDEE,IDBEE,IBEE,IDTKD,IKP,ICORDF,IH,NODF,IFLOW/10*4/
      DATA ICOORD,IDERVT,NOD/3*8/
      DATA IBTDB,IKM,IBT,IC,IVOLF,IDOF/6*16/,IKE/20/
C
      READ (5,*) NXE,NYE,N,IW,N3,IWF,NN,NGP,TOL,ITS
      READ (5,*) PERMX,PERMY
      READ (5,*) PHI,COH,PSI,E,V,CONS,P0
      READ (5,*) DTIM,INCS,THETA
      READ (5,*) (WIDTH(I),I=1,NXE+1)
      READ (5,*) (DEPTH(I),I=1,NYE+1)
      CALL READBC(NF,INF,NF2,NF3,NN,NODOF,LVEC)
      READ (5,*) LC,FACTOR
      READ (5,*) NL, (NO(I),VAL(I),I=1,NL)
C
      PI = ACOS(-1.)
      ITOT = IDOF + NODF
      IR = N* (IW+1)
      IR3 = N3* (IWF+1)
      NOELTS = NXE*NYE
C
      CALL NULVEC(DISPL,N)
      CALL NULVEC(OLDIS,N)
      CALL GAUSS(SAMP,ISAMP,NGP)
      CALL FMDRAD(DEE,IDEE,E,V)
      CALL NULL(KAY,IKAY,IT,IT)
      KAY(1,1) = PERMX
      KAY(2,2) = PERMY
C
```

```
      XOL(1) = 0.
      POGH2(1) = 0.
      NITS(1) = 0
C
      SNPH = SIN(PHI*PI/180.)
      DT = 4.* (1.+V)* (1.-2.*V)/ (E* (1.-2.*V+SNPH**2))
C
      DO 10 IY = 1,INCS
          ITER = 0
          CALL NULVEC(BDYLDS,N)
          CALL NULVEC(EVPT,NXE*NYE*IH*NGP*NGP)
          IF (IY.NE.1) GO TO 50
C
          CALL NULVEC(BK,IR)
          CALL NULVEC(BP,IR3)
C
          NM = 0
          DO 20 IP = 1,NXE
              DO 20 IQ = 1,NYE
                  NM = NM + 1
                  CALL G48UVP(IP,IQ,NXE,WIDTH,DEPTH,COORD,ICOORD,COORDF,
     +                       ICORDF,G,NF,INF,G3,NF3)
                  CALL NULL(KM,IKM,IDOF,IDOF)
                  CALL NULL(C,IC,IDOF,NODF)
                  CALL NULL(KP,IKP,NODF,NODF)
C
                  IG = 0
                  DO 30 I = 1,NGP
                      DO 30 J = 1,NGP
                          IG = IG + 1
C
C     SOLID CONTRIBUTION
C
                          CALL FMQUAD(DER,IDER,FUN,SAMP,ISAMP,I,J)
                          CALL GCOORD(FUN,COORD,ICOORD,NOD,IT,GC)
                          IF (IY.EQ.1) THEN
                              SX(IP,IQ,IG) = CONS
                              SY(IP,IQ,IG) = CONS
                              TXY(IP,IQ,IG) = 0.
                              SZ(IP,IQ,IG) = CONS
                              PORE(IP,IQ,IG) = P0
                          END IF
```

```
                         CALL MATMUL(DER,IDER,COORD,ICOORD,JAC,IJAC,IT,
      +                       NOD,IT)
                         CALL TWOBY2(JAC,IJAC,JAC1,IJAC1,DET)
                         CALL MATMUL(JAC1,IJAC1,DER,IDER,DERIV,IDERIV,
      +                       IT,IT,NOD)
                         CALL NULL(BEE,IBEE,IH,IDOF)
                         CALL FORMB(BEE,IBEE,DERIV,IDERIV,NOD)
                         CALL VOL2D(BEE,IBEE,VOL,NOD)
                         CALL MATMUL(DEE,IDEE,BEE,IBEE,DBEE,IDBEE,IH,
      +                       IH,IDOF)
                         CALL MATRAN(BT,IBT,BEE,IBEE,IH,IDOF)
                         CALL MATMUL(BT,IBT,DBEE,IDBEE,BTDB,IBTDB,IDOF,
      +                       IH,IDOF)
                         QUOT = DET*SAMP(I,2)*SAMP(J,2)
                         CALL MSMULT(BTDB,IBTDB,QUOT,IDOF,IDOF)
                         CALL MATADD(KM,IKM,BTDB,IBTDB,IDOF,IDOF)
C
C     FLUID CONTRIBUTION
C
                         CALL FORMLN(DERF,IDERF,FUNF,SAMP,ISAMP,I,J)
                         CALL MATMUL(DERF,IDERF,COORDF,ICORDF,JAC,IJAC,
      +                       IT,NODF,IT)
                         CALL TWOBY2(JAC,IJAC,JAC1,IJAC1,DET)
                         CALL MATMUL(JAC1,IJAC1,DERF,IDERF,DERIVF,
      +                       IDERVF,IT,IT,NODF)
                         CALL MATMUL(KAY,IKAY,DERIVF,IDERVF,KDERIV,
      +                       IKDERV,IT,IT,NODF)
                         CALL MATRAN(DERIVT,IDERVT,DERIVF,IDERVF,IT,
      +                       NODF)
                         CALL MATMUL(DERIVT,IDERVT,KDERIV,IKDERV,DTKD,
      +                       IDTKD,NODF,IT,NODF)
                         QUOT = DET*SAMP(I,2)*SAMP(J,2)
                         PROD = QUOT*DTIM
                         CALL MSMULT(DTKD,IDTKD,PROD,NODF,NODF)
                         CALL MATADD(KP,IKP,DTKD,IDTKD,NODF,NODF)
                         CALL VVMULT(VOL,FUNF,VOLF,IVOLF,IDOF,NODF)
                         CALL MSMULT(VOLF,IVOLF,QUOT,IDOF,NODF)
                         CALL MATADD(C,IC,VOLF,IVOLF,IDOF,NODF)
   30            CONTINUE
C
                 CALL FORMKE(KM,IKM,KP,IKP,C,IC,KE,IKE,IDOF,NODF,THETA)
                 CALL FORMKV(BK,KE,IKE,G,N,ITOT)
```

```
                        CALL FORMKV(BP,KP,IKP,G3,N3,NODF)
   20       CONTINUE
C
            IF (LC.EQ.0) THEN
                DO 40 I = 1,NL
                    BK(NO(I)) = BK(NO(I)) + 1.E20
   40           STORKB(I) = BK(NO(I))
            END IF
C
C      REDUCE LEFT HAND SIDE
C
            CALL BANRED(BK,N,IW)
   50       CONTINUE
C
C
            CALL NULVEC(ANS,N)
            MM = 0
            DO 60 K = 1,NN
                IF (NF(K,3).EQ.0) GO TO 60
                MM = MM + 1
                FLUID(MM) = DISPL(NF(K,3))
   60       CONTINUE
            CALL LINMUL(BP,FLUID,ANS2,N3,IWF)
            MM = 0
            DO 70 K = 1,NN
                IF (NF(K,3).EQ.0) GO TO 70
                MM = MM + 1
                ANS(NF(K,3)) = ANS2(MM)
   70       CONTINUE
C
   80       ITER = ITER + 1
C
            CALL VECADD(ANS,BDYLDS,LOADS,N)
C      CALL NULVEC(BDYLDS,N)
C
            IF (LC.EQ.0) THEN
                DO 90 I = 1,NL
   90           LOADS(NO(I)) = VAL(I)*FACTOR*STORKB(I)
            ELSE
                DO 100 I = 1,NL
  100           LOADS(NO(I)) = LOADS(NO(I)) + VAL(I)*FACTOR
            END IF
```

```
C
          CALL BACSUB(BK,LOADS,N,IW)
C
C      CONVERGENCE CHECK
C
          IZ = 1
          IF (ITER.EQ.1) THEN
              IZ = 0
          ELSE
              AK2 = .0
              DO 110 I = 1,NN
                  DO 110 J = 1,2
                      K = NF(I,J)
                      IF (K.EQ.0) GO TO 110
                      IF (ABS(LOADS(K)).GT.AK2) AK2 = ABS(LOADS(K))
 110              CONTINUE
              DO 120 I = 1,NN
                  DO 120 J = 1,2
                      K = NF(I,J)
                      IF (K.EQ.0) GO TO 120
                      AK1 = ABS((LOADS(K)-OLDIS(K))/AK2)
                      IF (AK1.GT.TOL) IZ = 0
 120              CONTINUE
              CALL VECCOP(LOADS,OLDIS,N)
          END IF
          IF (IZ.EQ.1 .OR. ITER.EQ.ITS) CALL NULVEC(BDYLDS,N)
C
C
C      CHECK ALL GAUSS POINTS
C
          FMAX = 0.
          NM = 0
          DO 130 IP = 1,NXE
              DO 130 IQ = 1,NYE
                  NM = NM + 1
                  CALL G48UVP(IP,IQ,NXE,WIDTH,DEPTH,COORD,ICOORD,COORDF,
      +                       ICORDF,G,NF,INF,G3,NF3)
                  DO 140 M = 1,IDOF
                      IF (G(M).EQ.0) ELD(M) = 0.0
 140              IF (G(M).NE.0) ELD(M) = LOADS(G(M))
                  CALL NULVEC(BLOAD,IDOF)
                  IG = 0
                  DO 150 I = 1,NGP
```

```
                        DO 150 J = 1,NGP
                           IG = IG + 1
                           IN = NGP*NGP*IH* (NM-1) + IH* (IG-1)
                           CALL FMDRAD(DEE,IDEE,E,V)
                           CALL FORMLN(DERF,IDERF,FUNF,SAMP,ISAMP,I,J)
                           CALL FMQUAD(DER,IDER,FUN,SAMP,ISAMP,I,J)
                           CALL MATMUL(DER,IDER,COORD,ICOORD,JAC,IJAC,IT,
         +                             NOD,IT)
                           CALL TWOBY2(JAC,IJAC,JAC1,IJAC1,DET)
                           CALL MATMUL(JAC1,IJAC1,DER,IDER,DERIV,IDERIV,
         +                             IT,IT,NOD)
                           CALL NULL(BEE,IBEE,IH,IDOF)
                           CALL FORMB(BEE,IBEE,DERIV,IDERIV,NOD)
                           CALL MATRAN(BT,IBT,BEE,IBEE,IH,IDOF)
                           CALL MVMULT(BEE,IBEE,ELD,IH,IDOF,EPS)
                           DO 160 K = 1,IH
  160                      EPS(K) = EPS(K) - EVPT(IN+K)
C
C      ELASTIC STRESS INCREMENT
C
                           CALL MVMULT(DEE,IDEE,EPS,IH,IH,SIGMA)
                           STRESS(1) = SIGMA(1) + SX(IP,IQ,IG)
                           STRESS(2) = SIGMA(2) + SY(IP,IQ,IG)
                           STRESS(3) = SIGMA(3) + TXY(IP,IQ,IG)
                           STRESS(4) = SIGMA(4) + SZ(IP,IQ,IG)
C
C      CHECK FOR FAILURE SURFACE VIOLATION
C
                           CALL INVAR(STRESS,SIGM,DSBAR,ALOD)
                           CALL MOCOUF(PHI,COH,SIGM,DSBAR,ALOD,F)
                           IF (F.GT.FMAX) FMAX = F
                           IF (IZ.EQ.1 .OR. ITER.EQ.ITS) GO TO 170
                           IF (F.LT.0.) GO TO 180
                           CALL MOCOUQ(PSI,DSBAR,ALOD,DQ1,DQ2,DQ3)
                           CALL FORMM(STRESS,M1,M2,M3)
                           DO 190 L = 1,IH
                              DO 190 M = 1,IH
  190                      FLOW(L,M) = F* (M1(L,M)*DQ1+M2(L,M)*DQ2+
         +                             M3(L,M)*DQ3)
                           CALL MVMULT(FLOW,IFLOW,STRESS,IH,IH,ERATE)
                           DO 200 K = 1,IH
                              EVP(K) = ERATE(K)*DT
```

```
   200                          EVPT(IN+K) = EVPT(IN+K) + EVP(K)
                               CALL MVMULT(DEE,IDEE,EVP,IH,IH,DEVP)
                               GO TO 210
   170                          CALL VECCOP(STRESS,DEVP,IH)
   210                          CALL MVMULT(BT,IBT,DEVP,IDOF,IH,ELOAD)
                               QUOT = DET*SAMP(I,2)*SAMP(J,2)
                               DO 220 K = 1,IDOF
   220                          BLOAD(K) = BLOAD(K) + ELOAD(K)*QUOT
   180                          IF (IZ.NE.1 .AND. ITER.NE.ITS) GO TO 150
C
C
C      UPDATE THE ELEMENT STRESSES
C
                               SX(IP,IQ,IG) = STRESS(1)
                               SY(IP,IQ,IG) = STRESS(2)
                               TXY(IP,IQ,IG) = STRESS(3)
                               SZ(IP,IQ,IG) = STRESS(4)
                               DPORE = 0.
                               DO 230 K = 1,NODF
                                  IF (G(K+IDOF).EQ.0) GO TO 230
                                  DPORE = DPORE + FUNF(K)*LOADS(G(K+IDOF))
   230                          CONTINUE
                               PORE(IP,IQ,IG) = PORE(IP,IQ,IG) + DPORE
   150             CONTINUE
C
C      COMPUTE TOTAL BODYLOADS VECTOR
C
                   DO 240 M = 1,IDOF
                      IF (G(M).EQ.0) GO TO 240
                      BDYLDS(G(M)) = BDYLDS(G(M)) + BLOAD(M)
   240             CONTINUE
   130     CONTINUE
C
         IF (IZ.NE.1 .AND. ITER.NE.ITS) GO TO 80
C
         CALL VECADD(DISPL,LOADS,DISPL,N)
C
C      OUTPUT
C
         MM = 0
         DO 250 M = 1,NN
            IF (NF(M,3).EQ.0) GO TO 250
```

```
                    MM = MM + 1
                    FLUID(MM) = DISPL(NF(M,3))
  250       CONTINUE
C
            XOL(IY+1) = -DISPL(1)
            POGH2(IY+1) = -IY*FACTOR
            NITS(IY+1) = ITER
            FM(IY+1) = FMAX
            NMAX = IY + 1
            IF (ITER.EQ.ITS) GO TO 260
   10 CONTINUE
C
  260 CONTINUE
      DO 270 I = 1,NMAX
  270 WRITE (6,'(2E12.4,I5,E12.4)') XOL(I),POGH2(I),NITS(I),FM(I)
      STOP
      END
C
      SUBROUTINE G48UVP(IP,IQ,NXE,WIDTH,DEPTH,COORD,ICOORD,COORDF,
     +                  ICORDF,G,NF,INF,G3,NF3)
C
C     THIS SUBROUTINE FORMS THE NODAL COORDINATES AND STEERING
C     VECTOR FOR A VARIABLE MESH OF 4-NODE/8-NODE
C     QUADRILATERAL ELEMENTS NUMBERING IN THE X-DIRECTION
C     (U,V,P  BIOT CONSOLIDATION)
C
      REAL COORD(ICOORD,*),COORDF(ICORDF,*),WIDTH(*),DEPTH(*)
      INTEGER G(*),NF(INF,*),NUM(8),G3(*),NF3(*)
      NUM(1) = IQ* (3*NXE+2) + 2*IP - 1
      NUM(2) = IQ* (3*NXE+2) + IP - NXE - 1
      NUM(3) = (IQ-1)* (3*NXE+2) + 2*IP - 1
      NUM(4) = NUM(3) + 1
      NUM(5) = NUM(4) + 1
      NUM(6) = NUM(2) + 1
      NUM(7) = NUM(1) + 2
      NUM(8) = NUM(1) + 1
      INC = 0
      DO 1 I = 1,8
          DO 1 J = 1,2
              INC = INC + 1
    1 G(INC) = NF(NUM(I),J)
      DO 2 I = 1,7,2
```

```
            INC = INC + 1
    2 G(INC) = NF(NUM(I),3)
      INC = 0
      DO 3 I = 1,7,2
            INC = INC + 1
    3 G3(INC) = NF3(NUM(I))
      COORD(1,1) = WIDTH(IP)
      COORD(2,1) = WIDTH(IP)
      COORD(3,1) = WIDTH(IP)
      COORDF(1,1) = WIDTH(IP)
      COORDF(2,1) = WIDTH(IP)
      COORD(5,1) = WIDTH(IP+1)
      COORD(6,1) = WIDTH(IP+1)
      COORD(7,1) = WIDTH(IP+1)
      COORDF(3,1) = WIDTH(IP+1)
      COORDF(4,1) = WIDTH(IP+1)
      COORD(4,1) = .5* (COORD(3,1)+COORD(5,1))
      COORD(8,1) = .5* (COORD(7,1)+COORD(1,1))
      COORD(1,2) = DEPTH(IQ+1)
      COORD(8,2) = DEPTH(IQ+1)
      COORD(7,2) = DEPTH(IQ+1)
      COORDF(1,2) = DEPTH(IQ+1)
      COORDF(4,2) = DEPTH(IQ+1)
      COORD(3,2) = DEPTH(IQ)
      COORD(4,2) = DEPTH(IQ)
      COORD(5,2) = DEPTH(IQ)
      COORDF(2,2) = DEPTH(IQ)
      COORDF(3,2) = DEPTH(IQ)
      COORD(2,2) = .5* (COORD(1,2)+COORD(3,2))
      COORD(6,2) = .5* (COORD(5,2)+COORD(7,2))
      RETURN
      END
C
      SUBROUTINE READBC(NF,INF,NF2,NF3,NN,NODOF,LVEC)
C
C     NODE FREEDOM ARRAY FOR
C      20 DOF B.C.ELEMENT
C
      INTEGER I,J,K,INF,NN,NODOF,KK,L,KKK
      INTEGER NF(INF,*),NF2(INF,*),NF3(*),LVEC(*)
      DO 10 I = 1,NN
          NF3(I) = 1
```

```
            DO 10 J = 1,NODOF
                NF(I,J) = 1
   10 IF (J.NE.3) NF2(I,J) = 1
C
C     FORMATIONS ARE
C     110,101,011,100,010,001,000
C
      DO 20 I = 1,7
          READ (5,*) K
          IF (K.NE.0) THEN
              READ (5,*) (LVEC(J),J=1,K)
              DO 30 KK = 1,K
                  L = LVEC(KK)
                  GO TO (1,2,3,4,5,6,7) I
    1             NF(L,3) = 0
                  NF3(L) = 0
                  GO TO 30
    2             NF(L,2) = 0
                  NF2(L,2) = 0
                  GO TO 30
    3             NF(L,1) = 0
                  NF2(L,1) = 0
                  GO TO 30
    4             NF(L,2) = 0
                  NF(L,3) = 0
                  NF2(L,2) = 0
                  NF3(L) = 0
                  GO TO 30
    5             NF(L,1) = 0
                  NF(L,3) = 0
                  NF2(L,1) = 0
                  NF3(L) = 0
                  GO TO 30
    6             NF(L,1) = 0
                  NF(L,2) = 0
                  NF2(L,1) = 0
                  NF2(L,2) = 0
                  GO TO 30
    7             NF(L,1) = 0
                  NF(L,2) = 0
                  NF(L,3) = 0
                  NF2(L,1) = 0
```

```
                      NF2(L,2) = 0
                      NF3(L) = 0
30            CONTINUE
        END IF
20 CONTINUE
   K = 1
   KK = 1
   KKK = 1
   DO 40 I = 1,NN
       IF (NF3(I).NE.0) THEN
            NF3(I) = KKK
            KKK = KKK + 1
       END IF
       DO 40 J = 1,NODOF
           IF (NF(I,J).NE.0) THEN
                NF(I,J) = K
                K = K + 1
           END IF
           IF (J.LT.3) THEN
                IF (NF2(I,J).NE.0) THEN
                     NF2(I,J) = KK
                     KK = KK + 1
                END IF
           END IF
40 CONTINUE
   RETURN
   END
```

5.4   Notes on the program

The listing consists of a main program MCPROG followed by two subroutines, G48UVP
and READBC. The program is a combination of the modified Program 9.2 MP92
described earlier in this report, and Program 6.0 for elasto–viscoplastic materials
published by Smith and Griffiths (1988). All the other subroutines are listed in the
Smith and Griffiths text with the exception of FORMKE which was described at the end
of program MP92.

This program has many features in common with those already published, however
the following points are worthy of emphasis:

1. Unlike MP92, this program computes the fluid contribution to nodal loading at

the *global* level. This necessitates the formation of a global fluid stiffness matrix $\mathbf{K_p}$ (BP) and extraction of the global pore pressures (FLUID) from the global 'displacements' vector (DISPL). The global matrix/vector product is computed by subroutine LINMUL. This approach requires more memory than the element–by–element technique, but avoids the need to scan through all the elements recovering or reforming their $\mathbf{k_p}$ matrices.

2. Since two global matrices are formed by this program (BK and BP), two separate assembly systems must run in parallel, hence the special geometry subroutine G48UVP which generates steering vectors (G and G3) for both the entire set of freedoms and those corresponding to the fluid freedoms only.

3. The program allows either load control (LC = 1) or displacement control (LC = 0) which is set by the data.

4. The program assumes a constant loading (or displacement) rate although quite general loading rates could be easily incorporated by including subroutine LFUNC from program MC92.

5. Subroutine READBC saves on the amount of data that must be read in to define the boundary conditions. For Biot-type problems, all mid–side nodes have a 'restraint' in the sense that the pore pressure is not explicitly computed at those locations. This can lead to rather long and tedious data preparation. READBC reads in the node numbers at which particular restraints apply. There are seven possible restraint types and these are covered in a fixed sequence (see comment lines in READBC).

For problems with more complicated boundaries, the user must replace the geometry subroutine G48UVP, which is only suitable for rectangular geometries, by some other means of generating the nodal coordinates COORD and the 'steering' vectors G and G3.

5.5   Sample data

The following data represents the 'fast' loading case where $dD_f/dt = 5 \times 10^6$

```
                         Mesh Data
        NXE  NYE   N   IW  N3  IWF  NN  NGP   TOL   ITS
         2    2   36   21   4    3   21   2   0.001  250
                       Element data
     PERMX  PERMY  PHI  COH  PSI   E     V   CONS  PO
      1.     1.    30.  0.   0.   1.e4  0.25  -10.  0.
                  Time integration data
                  DTIM      ISTEP   THETA
               0.0000001     41      1.0

            Geometry data -- WIDTH , DEPTH
                     0.   0.5   1.0
                     0.  -0.5  -1.0

            Node freedom data -- READBC
                     11                     type (110)
      2  3  4  5   7   8  10  12  13  15  16
                      1                      type (101)
                     19
                      1                      type (011)
                      9
                      3                      type (100)
                 18  20  21
                      3                      type (010)
                  1   6  14
                      1                      type (001)
                     17
                      0                      type (000)
              Loading data -- Control
                   LC  FACTOR
                    1   -0.5
            Loading data -- Nodal weightings
         5
         1  0.08333333  3  0.33333333  5  0.16666667
         7  0.33333333  9  0.08333333
```

The majority of variables read in as data have been encountered before, however
those that have not are summarised in Table 5.2.

Table 5.2: **Meaning of some data variables**

N3	Number of active pore pressure freedoms
IWF	Band–width of pore pressure freedoms
TOL	Convergence tolerance
ITS	Iteration ceiling at each time step
PHI	Soil friction angle ϕ'
COH	Soil cohesion c'
PSI	Soil dilation angle ψ'
CONS	Initial stresses $\sigma_x, \sigma_y, \sigma_z$
P0	Initial pore pressure
INCS	Maximum number of time steps
LC	= 1, load control; = 0, displacement control
FACTOR	Load (or displacement) increment at each time step

5.6   Sample output

The following output was produced by the listed program and data. The four columns of results represent the axial displacement δ_1, the deviator stress D_f, the number of viscoplastic time–steps (or iterations) to reach convergence at each Biot time step, and the maximum value of the Mohr–Coulomb failure function F at convergence.

δ_1	D_f	ITER	F_{max}
0.0000E-00	0.0000E+00	0	0.0000E+00
0.3342E-04	0.5000E+00	3	0.0000E+00
0.6689E-04	0.1000E+01	3	0.0000E+00
0.1004E-03	0.1500E+01	3	0.0000E+00
0.1340E-03	0.2000E+01	3	0.0000E+00
0.1676E-03	0.2500E+01	3	0.0000E+00
0.2013E-03	0.3000E+01	3	0.0000E+00
0.2351E-03	0.3500E+01	3	0.0000E+00
0.2689E-03	0.4000E+01	3	0.0000E+00
0.3027E-03	0.4500E+01	3	0.0000E+00
0.3367E-03	0.5000E+01	3	0.0000E+00
0.3706E-03	0.5500E+01	3	0.0000E+00
0.4047E-03	0.6000E+01	3	0.0000E+00
0.4388E-03	0.6500E+01	3	0.0000E+00
0.4729E-03	0.7000E+01	3	0.0000E+00
0.5071E-03	0.7500E+01	3	0.0000E+00
0.5414E-03	0.8000E+01	3	0.0000E+00
0.5757E-03	0.8500E+01	3	0.0000E+00
0.6123E-03	0.9000E+01	9	0.4435E-03
0.6673E-03	0.9500E+01	14	0.8788E-03
0.4337E-02	0.1000E+02	250	0.1677E+00

Failure of the element is indicated by:

- Sudden increase in displacement δ_1.

- ITER reaches the iteration ceiling of 250.

- $F_{max} > 0$ suggesting a lack of convergence.

Chapter 6

Further applications

This chapter is devoted to three examples in which the viscoplastic/Biot algorithm is applied to more advanced boundary value problems of interest to Civil Engineers.

6.1   Thick walled cylinder

The analysis is of a thick-walled cylinder of saturated soil in plane strain subjected to an internal pressure which increases linearly with time. The problem is to estimate the ultimate internal pressure that the cylinder can sustain as a function of the *rate* of loading. Results for this type of problem were first reported by Small *et al* (1976) and has applications relating to in–situ pressuremeter testing.

A typical finite element mesh is shown in figure 6.1 for a cylinder in which the outer radius is twice the inner radius. The mesh consists of 8-node elements for the solid phase and 4-node elements for the fluid phase (Zienkiewicz 1977). Four Gauss- points per element were used for all integration.

The soil was initially stress-free and given the properties indicated in Table 6.1.

Table 6.1: **Soil properties for thick–walled cylinder analysis**

E'	$=$	200 kPa	ν'	$=$	0
ϕ'	$=$	$30°$	c'	$=$	1 kPa
ψ'	$=$	0	k/γ_w	$=$	1 m^4/(kNs)

A dimensionless time factor is introduced, given by:

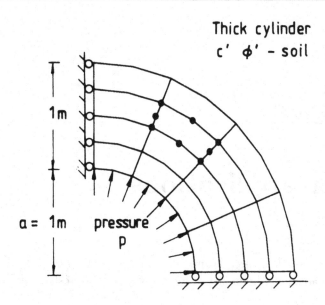

Figure 6.1: **Mesh used for thick–walled cylinder analysis**

$$T_v = \frac{c_v t}{a^2} \tag{6.1}$$

where a is the internal radius of the cylinder (in this case equal to unity), and t represents time. The coefficient of consolidation c_v is defined:

$$c_v = \frac{k(1 - \nu')E'}{\gamma_w(1 - 2\nu')(1 + \nu')} \tag{6.2}$$

hence $c_v = 200\text{m}^2/\text{s}$ in this case.

The rate of increase of internal pressure acting on the cylinder is expressed in the following non-dimensional form:

$$\omega = \frac{d(p/c')}{dT_v} \tag{6.3}$$

where p is the time dependent internal pressure on the cylinder.

The behaviour of the soil during 'fast' loading approximates to undrained conditions. With the dilation angle equal to zero, no volume change will occur during shear and the mean effective stress will remain constant and equal to zero during loading. For the undrained case, the soil 'properties' shown in Table 6.2 will operate.

Table 6.2: **Undrained soil properties**

E_u	$=$	$1.5E'/(1+\nu') = 300\text{kPa}$	ν_u	$\approx$	0.5
ϕ_u	$=$	0	c_u	$=$	$c'\cos\phi'$

The failure load quoted by Small *et al* for the drained problem was:

$$p/c' = 1.02 \tag{6.4}$$

In the undrained case, Prager and Hodge (1968) gave $p/c_u = 2\ln 2$, which for a Mohr-Coulomb material with $\phi' = 30°$ is equivalent to:

$$p/c' = 2\ln 2\cos 30° = 1.20 \tag{6.5}$$

The non-dimensionalised pressure-deflection curves computed using the present method are given in Figure 6.2 for three different loading rates given by $\omega = 16$ (fast), $\omega = 4$ (intermediate) and $\omega = 0.1$ (slow).

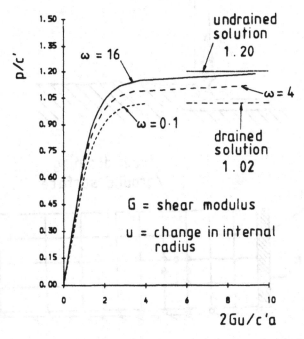

Figure 6.2: **Stress vs. Displacement for different loading rates**

6.2 Passive earth pressure

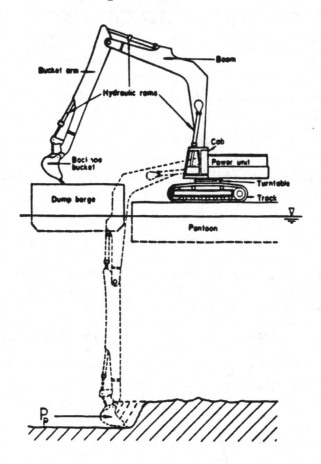

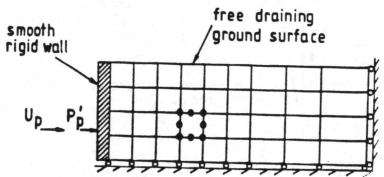

Figure 6.3: **Dredging and passive earth pressure analysis**

The transient analyses were next applied to a problem of passive earth pressure. The practical applications of such a study relate to the dredging process, whereby underwater soils are removed by excavators in order to maintain sufficient depth in ports, harbours, rivers and canals.

In this section, the effect of wall speed on passive resistance of non-dilative soils is considered. The plane strain mesh geometry is shown in Figure 6.3. The simple boundaries of the problem removes the shear concentration that would normally be observed at the base of the wall, but does not detract from the analysis in the case of a smooth wall.

The excavation blade is idealised as a rigid vertical smooth wall translated horizontally into the soil bed. All boundaries of the mesh are impermeable except the top surface at which drainage can occur.

The soil is given the properties indicated in Table 6.3.

Table 6.3: **Soil properties used in earth pressure analysis**

E' = 1×10^4 kPa	ν' = 0.25	
c' = 0 kPa	ϕ' = $30°, (K_p = 3)$	
ψ' = 0	γ' = 10 kN/m^3	
K_o = 0.5	k/γ_w = 1×10^{-5} m^4/(kNs)	
c_v = 0.12 m^2/s		

where γ' is the submerged unit weight and K_o is the coefficient of earth pressure at rest. The wall height equals 1 metre.

A dimensionless dredging rate L_R is used to represent the (constant) speed of the wall as it moves into the soil. This quantity is defined:

$$L_R = \frac{d(x/H)}{dT_d} \tag{6.6}$$

with the time factor T_d given by:

$$T_d = \frac{c_v t}{H^2} \tag{6.7}$$

The height of the wall is H, and x represents the distance moved by the wall after time t. The effective soil force P_p' against the wall was computed by integrating the horizontal effective stresses in the column of elements adjacent to the wall. These forces are plotted in dimensionless form in Figure 6.4 for three rates of wall displacement. In each case, the total wall displacement was achieved numerically using 60 equal increments. The calculation time step was 0.001. 0.1 and 10 seconds respectively for the 'fast', 'intermediate' and 'slow' cases.

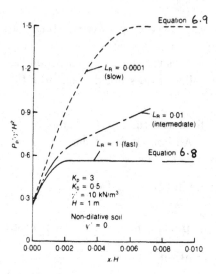

Figure 6.4: **Force vs. Displacement for different displacement rates**

At the fast dredging rate ($L_R = 1.0$), the soil behind the wall behaves in an undrained manner, whereas at the slow dredging rate ($L_R = 0.0001$), drained conditions are approximated. An intermediate response is observed at the compromise dredging rate ($L_R = 0.01$).

In the undrained case, the analytical solution (Griffiths 1985, Li 1988) for the effective passive force is given by :

$$P'_p = \frac{1}{2}\gamma' H^2 \frac{K_p(K_o + 1)}{K_p + 1} \tag{6.8}$$

In the drained case, the simple Rankine solution is obtained:

$$P'_p = \frac{1}{2}\gamma' H^2 K_p \tag{6.9}$$

The computed values in Figure 6.4 are seen to agree closely with eqns. 6.8 and 6.9.

As the soil model used in these analyses was non-dilative, the slow moving wall (drained) led to greater passive resistance than the fast moving wall (undrained) in which compressive pore pressures were generated in the elastic phase of the stress-strain behaviour.

Figure 6.5 shows the effect of the dilation angle on the effective force behind the wall for the 'fast' and 'slow' cases. In the 'fast' case, the effective force continues to rise well past the drained Rankine solution. In the 'slow' case, the dilation angle makes virtually no difference to the solution, and the result converges on the drained solution given by eqn. 6.9.

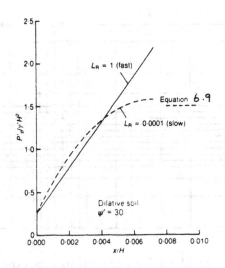

Figure 6.5: **Force vs. Displacement for different dilation angles**

In the present work, the dilation angle is constant and this is a deficiency of the simple model. In 'real' soils, continuous shearing will reduce the rate of dilation until the critical state is reached, at which point soil continues to shear at constant volume and constant pore pressure (see e.g. Seed and Lee 1967) .

6.3   Excavation

This section describes some analysis in which the stability of a vertical cut in a saturated c', ϕ' soil is shown to be affected by the *rate* of excavation.

The excavation was simulated by the removal of horizontal rows of elements in the mesh. Forces generated by this action were applied to the new mesh boundary. The boundary forces at the ith stage of an excavation are given by:

$$\mathbf{F}_i = \int_{V_i^e} \mathbf{B}^\mathbf{T} \, \sigma_{i-1} \, \mathrm{d}(\mathrm{vol}) - \int_{V_i^e} \mathbf{N}^\mathbf{T} \, \gamma \, \mathrm{d}(\mathrm{vol}) \qquad (6.10)$$

The first term in eqn. 6.10 is the nodal internal resisting force vector due to the stresses that existed in the removed elements, and the second term represents the reversal of the nodal body-forces of those elements assuming that γ is acting in the negative direction (see e.g. Ghaboussi and Pecknold 1984 , Holt 1991).

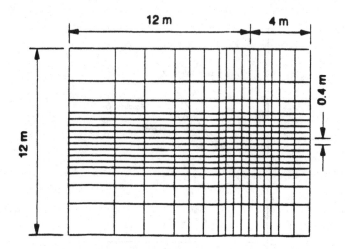

Figure 6.6: **Mesh for excavation analyses**

Figure 6.6 shows a typical finite element mesh for the excavation analyses. The analysis is in plane strain so the mesh will represent half the eventual excavation. The soil properties are given in Table 6.4.

Table 6.4: **Soil properties used in excavation analysis**

E'	=	1.0×10^5 kPa	ψ'	=	$0.0°$
ν'	=	0.43	γ'	=	20.0 kN/m^3
ϕ'	=	$15°$	K_o	=	0.75
c'	=	$16°$	k/γ_w	=	1.0 m^4/(kNs)

For the given values of c' and ϕ', the drained critical depth from Taylor's (1948) charts is 4.0m. Excavation of a 4m wide strip was carried out 'slowly' until failure occurred. The computed result indicated a critical depth at 4.4m, some 9.1% higher than Taylor's value. The difference was thought to be due to the rather crude mesh of Figure 6.6 in which each excavation was limited to a minimum of 0.4m.

The previous result was obtained with a 'slow' excavation leading to a drained solution. In order to investigate rate-effects on excavation stability, the next set of results were obtained by keeping a constant excavation rate of 1 m/s but systematically reducing the permeability coefficient K $(= k/\gamma_w)$ by an order of magnitude at a time in the range:

$$1.0 \times 10^{-10} \leq K_{x,y} \leq 1.0 \times 10^{-1} \qquad m^4/(kNs) \qquad (6.11)$$

As the permeability was reduced, the excavation tended towards a 'fast' rate and hence an undrained solution.

It was found that the lower the permeability of the soil the higher the value of the critical depth obtained due to the retention of the pore fluid following unloading effects during excavation. The upper limit of the critical depth corresponding to undrained conditions was found to be 5.6m for values of the permeability coefficient of 1.0×10^{-7} m^4/(kNs) or less.

To investigate more closely the transient region between the upper and lower limits ('drained' or 'undrained'), a considerably finer mesh was employed together with a smaller step size on the permeability coefficient (Holt 1991). As shown in Figure 6.7, the upper limit for the critical depth was unchanged at 5.6m, however the lower limit was reduced to 4.2m, now less than 5% in 'error' with Taylor's analytical solution.

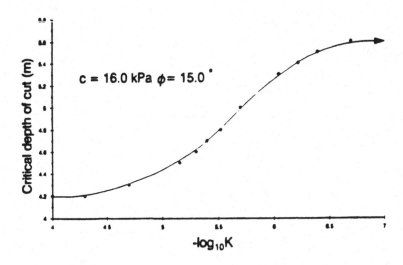

Figure 6.7: **Critical height vs. permeability (constant excavation rate)**

6.4   Concluding remarks

This chapter has described three examples of transient analysis in geomechanics in which the strength of a saturated soil mass has depended on the *rate* of loading. The algorithm involved combining an *incremental* version of Biot's equations for a porous elastic medium with a simple elastic-perfectly plastic Mohr-Coulomb constitutive law. In all cases, the drained and undrained limits were retrieved as special cases of the transient analysis and comparison made with analytical solutions when available.

It should be emphasised that the results presented herein were based on a simple elasto-plastic soil model which assumed no plastic volume change at failure. The resulting undrained stress paths approximate to 'medium' dense materials; however if the soil under investigation exhibited significant dilative or contractive tendencies, the 'simple' model would not be suitable and a more sophisticated soil model would be required. Of particular interest would be a soil which exhibited contractive tendencies during shear. Such a material could experience a rapid loss of strength during undrained loading. In the context of excavation analysis for example, this kind of behaviour might lead to the undrained case being *more* critical than the drained case. Such behaviour was indicated in examples of slope stability by Hicks and Wong (1988) , and studies are continuing in this area.

Bibliography

[1] M.A. Biot. General theory of three-dimensional consolidation. *J Appl Phys*, 12:155–164, 1941.

[2] O.C. Zienkiewicz. *The finite element method*. McGraw Hill, London, New York, 3rd edition, 1977.

[3] M.A. Hicks. *Numerically modelling the stress-strain behaviour of soils*. PhD thesis, Department of Engineering, University of Manchester, 1990.

[4] R.S. Sandhu. Finite element analysis of coupled deformation and fluid flow in porous media. In J.B. Martins, editor, *Numerical Methods in Geomechanics*, pages 203–228. D. Reidel Publishing Company, Dordrecht, Holland, 1981.

[5] D.V. Griffiths, M.A. Hicks, and C.O.Li. Transient passive earth pressure analyses. *Géotechnique*, 41(4):615–620, 1991.

[6] R.L. Schiffman. Field applications of soil consolidation, time–dependent loading and variable permeability. Technical Report 248, Highway Research Board, Washington, U.S.A., 1960.

[7] I.M. Smith and D.V. Griffiths. *Programming the Finite Element Method*. John Wiley and Sons, Chichester, New York, 2nd edition, 1988.

[8] D.C. Drucker and W. Prager. Soil mechanics and plastic analysis in limit design. *Q Appl Mech*, 10:157–165, 1952.

[9] D.C. Drucker, R.E. Gibson, and D.J. Henkel. Soil mechanics and work-hardening theories of plasticity. *Trans ASCE*, 122:338–346, 1957.

[10] C. Humpheson. *Finite element analysis of elasto/viscoplastic soils*. PhD thesis, Department of Civil Engineering, University of Wales at Swansea, 1976.

[11] O.C. Zienkiewicz, V.A. Norris, L.A. Winnicki, D.J. Naylor, and R.W. Lewis. A unified approach to the soil mechanics of offshore foundations. In *Numerical methods in offshore engineering*, pages 361–412. John Wiley and Sons, Chichester, New York, 1978.

[12] D.V. Griffiths. Some theoretical observations on conical failure criteria in principal stress space. *Int J Solids Struct*, 22(5):553–565, 1986.

[13] A.W. Bishop. The strength of soils as engineering materials: 6th Rankine Lecture. *Géotechnique*, 16(2):91–130, 1966.

[14] P.V. Lade and J.M. Duncan. Elasto-plastic stress-strain theory for cohesionless soils. *J Geotech Eng, ASCE*, 101(GT10):1037–1053, 1975.

[15] H. Matsuoka and T. Nakai. Stress-deformation and strength characteristics of soil under three different principal stresses. *Proc Jap Soc Civ Eng*, (232):59–70, 1974.

[16] T.W. Lambe and R.V. Whitman. *Soil Mechanics*, page 206. John Wiley and Sons, New York, 1969.

[17] O.C. Zienkiewicz and I.C. Cormeau. Viscoplasticity, plasticity and creep in elastic solids. a unified approach. *Int J Numer Methods Eng*, 8:821–845, 1974.

[18] I.C. Cormeau. Numerical stability in quasi–static elasto–viscoplasticity. *Int J Numer Methods Eng*, 9(1):109–127, 1975.

[19] D.V. Griffiths. *Finite element analyses of walls, footings and slopes*. PhD thesis, Department of Engineering, University of Manchester, 1980.

[20] D.V. Griffiths. The effect of pore fluid compressibility on failure loads in elasto-plastic soils. *Int J Numer Anal Methods Geomech*, 9:253–259, 1985.

[21] J.C. Small, J.R. Booker, and E.H. Davis. Elasto-plastic consolidation of soil. *Int J Solids Struct*, 12:431–448, 1976.

[22] W. Prager and P.G. Hodge. *Theory of perfectly plastic solids*. Dover Publications Inc., 1968.

[23] C.O. Li. *Finite element analyses of seepage and stability problems in geomechanics*. PhD thesis, Department of Engineering, University of Manchester, 1988.

[24] H.B. Seed and K.L. Lee. Undrained strength characteristics of cohesionless soils. *J Soil Mech Found Div, ASCE*, 93(SM6):333–360, 1967.

[25] J. Ghaboussi and D.A. Pecknold. Incremental finite element analysis of geometrically altered structures. *Int J Numer Methods Eng*, 20:2151–2164, 1984.

[26] D.A. Holt. Transient analysis of excavations in soil. Master's thesis, Department of Engineering, University of Manchester, 1991.

[27] D.W. Taylor. *Fundamentals of soil mechanics*. John Wiley and Sons, Chichester, New York, 1948.

[28] M.A. Hicks and S.W. Wong. Static liquefaction of loose slopes. In G. Swoboda, editor, *Proc 6th Int Conf Numer Methods Geomech*, pages 1361–1368. A.A. Balkema, Rotterdam, 1988.

Printed in the United States
By Bookmasters